Evolution!

Facts and Fallacies

Evolution!
Facts and Fallacies

Edited by
J. William Schopf

Center for the Study of Evolution and the Origin of Life
Department of Earth and Space Sciences
University of California
Los Angeles, California

Academic Press

San Diego London Boston New York Sydney Tokyo Toronto

A contribution of the IGPP Center for the Study of Evolution and the Origin of Life (CSEOL), University of California, Los Angeles.

Photos for the biographical sketches courtesy of Richard Mantonya

This book is printed on acid-free paper. ∞

Academic Press
a division of Harcourt Brace & Company
525 B Street, Suite 1900, San Diego, California 92101-4495, USA
http://www.apnet.com

Academic Press
24-28 Oval Road, London NW1 7DX, UK
http://www.hbuk.co.uk/ap/

Library of Congress Card Catalog Number: 98-84494

International Standard Book Number: 0-12-628860-7

PRINTED IN THE UNITED STATES OF AMERICA
98 99 00 01 02 03 MM 9 8 7 6 5 4 3 2 1

CONTENTS

CHAPTER 3
MISSING LINKS IN THE HISTORY OF LIFE
Charles R. Marshall

CHAPTER 4
BEYOND REASON: SCIENCE IN THE MASS MEDIA
Jere H. Lipps

CHAPTER 5
BREAKTHROUGH DISCOVERIES
J. William Schopf

CHAPTER 6
ARE WE ALONE IN THE COSMOS?
Tobias C. Owen

PREFACE

More than 3500 students, faculty, and other members of the UCLA community attended the "Evolution! Facts and Fallacies" symposium on March 15, 1997, convened by the IGPP Center for the Study of Evolution and the Origin of Life (CSEOL) at the University of California, Los Angeles. This volume makes accessible the proceedings of that symposium.

The symposium was held in honor of the renowned scientist and communicator **Carl Sagan**, David Duncan Professor of Astronomy and Space Sciences at Cornell University, who prior to his death from a bone marrow disorder on December 20, 1996, had planned to speak at the symposium. Carl had special rapport with people worldwide, exemplified by his 1980 *Cosmos,* which remains the most watched series in the history of public television. In his last published volume, *The Demon-Haunted World* (Random House, 1995), he railed against the American embrace of pseudoscience, a theme he was to rejoin at the symposium in a discussion he titled "The Conceit of Being Central to the Purpose of the Universe."

In Carl's words, "We make our world significant by the courage of our questions and the depth of our answers." The goal of the authors of this volume is to address the central questions of evolution and the relation of science to society in a way that would make him proud.

Biographical sketches of the authors of the six chapters of the book follow.

The late Carl Sagan (Cornell University, Ithaca, New York) – scientist, teacher, communicator, friend.

BIOGRAPHICAL SKETCHES

Chapter 1: The Evolution of Life

STEPHEN JAY GOULD

A native and now part-time resident of New York City, Stephen Jay Gould received his undergraduate training in geology at Antioch College, Ohio, and his Ph.D. from Columbia University. After receiving his doctorate in 1967, he joined the faculty at Harvard University, where he is Professor of Geology, Alexander Agassiz Professor of Zoology, and Curator of Invertebrate Paleontology at the Museum of Comparative Zoology and holds appointments in the Departments of Geology, Biology, and the History of Science. World-renowned as a scientist, scholar, author, and teacher, Professor Gould is the recipient of 40 honorary academic degrees, literary prizes for *This View of Life* (1979), *The Panda's Thumb* (1980), *The Mismeasure of Man* (1981), *Hen's Teeth and Horse's Toes* (1983), and *Wonderful Life* (1989), and more than 40 other national and international medals and awards,

Stephen Jay Gould (Harvard University, Cambridge, Massachusetts).

including the UCLA Medal (1992). Past-President of the Paleontological Society and the Society for the Study of Evolution and a member of the National Academy of Sciences, the American Academy of Arts and Sciences, the Royal Society of Edinburgh, and the Linnaean Society of London, in 1981 he was among the first recipients of a MacArthur Foundation "Genius Award" Prize Fellowship.

Chapter 2: Dating the Timeline of Life's History

JON P. DAVIDSON

Born in Welwyn Garden City, Herts, England, Jon Paul Davidson received a First Class Honours Degree in Geological Sciences from the University of Durham and his Ph.D. in 1984 from the Department of Earth Sciences at the University of Leeds. Recipient of a Fullbright Travel Award, he served as Visiting Professor in the Department of Geological Sciences and as Research Associate at the Institute for the Study of Earth and Man at Southern Methodist University from 1984 to 1987, and in the following year was Visiting Professor in the Department of Geological Sciences at the University of Michigan. In 1988 he joined the faculty of the UCLA Department of Earth and Space Sciences, where he has served as Vice Chair of the Department and is now Associate Professor of Geology and Geochemistry and Graduate Advisor of the Geochemistry Program. His numerous scientific publications have centered on use of isotopes in rocks and minerals as tracers of geologic events that shaped the volcanic "Ring of Fire" rimming the Pacific Ocean from Kamchatka, Russia, to the Andes of southern Chile. An exemplary scientist and acclaimed lecturer and mentor, Professor Davidson was a 1994 recipient of the UCLA Luckman Distinguished Teaching Award.

Jon P. Davidson (University of California, Los Angeles).

Chapter 3: Missing Links in the History of Life

CHARLES R. MARSHALL

Born in Canberra, Australia, Charles Richard Marshall received his undergraduate training in zoology at the Australian National University and his Ph.D. in 1989 from the Committee of Evolutionary Biology at the University of Chicago. After serving as a National Institutes of Health Postdoctoral Fellow at Indiana University, Bloomington, he joined the faculty at UCLA, where he is a member of the Department of Earth and Space Sciences, the Molecular Biology Institute, and the Institute of Geophysics and Planetary Physics. Recipient of a National Science Foundation Young Investigator Award and a Research Associate of the Los Angeles County Museum of Natural History, Professor Marshall is not only a superb scientist but a gifted teacher, selected in 1995 as a Paleontological Society Distinguished Lecturer. Convenor of the 1994 symposium in this series *Evolution and the Molecular Revolution* and co-editor of the resultant volume, Dr. Marshall is in the vanguard of a new breed of evolutionary biologists with in-depth understanding of mathematics and molecular biology in addition to paleontology and the workings of evolution, expertise he has applied to studies of the incompleteness of the fossil record, extinctions in the history of life, and biochemical evolution and systematics.

Charles R. Marshall (University of California, Los Angeles).

Chapter 4: Beyond Reason: Science in the Mass Media

JERE H. LIPPS

A native Californian, Jere H. Lipps was born in Los Angeles, was educated at UCLA (B.A., 1962; Ph.D., 1966), and has held professorial appointments at the University of California, Davis (1967–1988), and, since 1988, the University of California, Berkeley, where he is Director of the Museum of Paleontology and former Chair of the Department of Integrative Biology. A micropaleontologist, ecologist, and geologist who has served as a Visiting Biologist at Eniwetok, Bimini, and Papua New Guinea and a participant in the Ross Ice Shelf and deep-sea drilling projects, has focused on evolution in marine ecosystems, particularly shelled protozoans. A Fellow of the American Association for the Advancement of Science and the California Academy of Sciences, Professor Lipps has served as a Visiting Scientist at the British Museum, Guest Professor at Aarhus University, Denmark, and Paleontological Society Dististinguished Lecturer. He has been awarded the Antarctic Medal of the United States, conferred the distinction of having Lipps Island off the Antarctic coast named in his honor, and in 1993 was recipient of the W. Storrs Cole Award in Micropaleontology of the Geological Society of America. He is current President of the Paleontological Society.

Jere H. Lipps (University of California, Berkeley).

Chapter 5: Breakthrough Discoveries

J. WILLIAM SCHOPF

Director of the Center for the Study of Evolution and the Origin of Life (CSEOL) and editor of this volume, James William Schopf is a member of the UCLA Department of Earth and Space Sciences, Molecular Biology Institute, and Institute of Geophysics and Planetary Physics. An undergraduate geology major at Oberlin College, Ohio, in 1968 he received his Ph.D. in biology from Harvard University. He has edited or co-edited four volumes from previous CSEOL-sponsored symposia and two prize-winning monographs on the earliest history of life on Earth and has been honored as a Distinguished Teacher, Faculty Research Lecturer, and recipient of the UCLA Gold Shield Prize for Academic Excellence. The first foreign member elected to the Scientific Presidium of the Russian Academy of Science's A.N. Bach Institute of Biochemistry and a member of National Academy of Sciences, the American Philosophical Society, and the American Academy of Arts and Sciences, Professor Schopf has been awarded two Guggenheim Fellowships and medals by the National Academy of Sciences, the National Science Board, and the International Society for the Study of the Origin of Life. He is discoverer of the oldest fossil organisms known and has carried out geological studies in Africa, Asia, Australia, Europe, and North and South America.

J. William Schopf (University of California, Los Angeles).

Chapter 6: Are We Alone in the Cosmos?

TOBIAS C. OWEN

Tobias C. Owen received his higher education in liberal arts, physics, and astronomy at the University of Chicago (B.A., B.S., M.S.) and the University of Arizona (Ph.D., 1965). Before assuming his current position of Astronomer at the University of Hawaii in 1990, he served on the faculty of the State University of New York at Stony Brook (1970–1990) and as a Research Physicist at the Astro Sciences Center of IIT Research Institute (1964–70), a Visiting Professor at the California Institute of Technology (1970), and Astronome Titulaire at the Paris Observatory (1980). He has participated in missions exploring Mars, Jupiter, Saturn, and the outer planets and is recipient of four NASA Group Achievement Awards and the NASA Medal for Exceptional Scientific Achievement, the Alumni Professional Achievement Award of the University of Chicago, and the Newcomb Cleveland Prize of the American Association for the Advancement of Science. Author of *The Search for Life in the Universe* (with D. Goldsmith) and *The Planetary System* (with D. Morrison) and a Fellow of the American Geophysical Union and the American Association for the Advancement of Science, he has focused his interests on the physics and chemistry of the solar system and the origin and cosmic distribution of life.

Tobias C. Owen (University of Hawaii, Honolulu).

CHAPTER 1

The Evolution of Life

Stephen Jay Gould[1]

INTRODUCTION

Some creators announce their inventions with grand èclat. God proclaimed, "Fiat lux," and then flooded his new universe with brightness. Others bring forth great discoveries in a modest guise, as did Charles Darwin in defining his new mechanism of evolutionary causality in 1859: "I have called this principle, by which each slight variation, if useful, is preserved, by the term **Natural Selection**."

Natural selection is an immensely powerful yet beautifully simple theory that has held up remarkably well, under intense and unrelenting scrutiny and testing, for 140 years. In essence, natural selection locates the mechanism of evolutionary change in a "struggle" among organisms for reproductive success, leading to improved fit of populations to changing environments. (Struggle is often a metaphorical description and need not be viewed as overt combat, guns blazing. Tactics for reproductive success include a variety of nonmartial activities such as earlier and more frequent mating or better cooperation with partners in raising offspring.) Natural selection is therefore a principle of local adaptation, not of general advance or progress (see Figure 1.1).

Yet powerful though the principle may be, natural selection is not the only cause of evolutionary change (and may, in many cases, be overshadowed by other forces). This point needs emphasis because the standard misapplication of evolutionary theory assumes that biological explanation may be equated with devising accounts, often speculative and conjectural in practice, about the adaptive value of any given feature in its original environment (human aggression as good for hunting, music and religion as good for tribal cohesion, for example). Darwin himself strongly emphasized the multifactorial nature of evolutionary change and warned against too exclusive a reliance on natural selection, by placing the following statement in a maximally conspicuous place at the very end of his introduction: "I am convinced that Natural Selection has been the most important, but not the exclusive, means of modification."

[1]Museum of Comparative Zoology, Harvard University, Cambridge, MA 02138. This essay is adapted and reprinted with permission. Copyright ©1994 by Scientific American, Inc. All rights reserved.

FIGURE 1.1   This slab containing specimens of *Pteridinium* from Namibia shows a prominent organism from the Earth's first multicellular fauna, called Ediacaran, which appeared some 600 million years ago. The Ediacaran animals died out before the Cambrian explosion of modern life. Though these thin quilted sheetlike organisms may be ancestral to some modern forms they also may represent a separate and ultimately failed experiment in multicellular life. The history of life tends to move in quick and quirky episodes, rather than by gradual improvement.

UNDERSTANDING LIFE'S PATHWAY

Natural selection is not fully sufficient to explain evolutionary change for two major reasons. First, many other causes are powerful, particularly at levels of biological organization both above and below the traditional Darwinian focus on organisms and their struggles for reproductive success. At the lowest level of substitution in individual base pairs of **DNA**, change is often effectively neutral and therefore random. At higher levels, involving entire species or faunas, **punctuated equilibrium** can produce evolutionary trends by selection of species based on their rates of origin and extirpation, whereas mass extinctions wipe out substantial parts of biotas for reasons unrelated to adaptive struggles of constituent species in "normal" times between such events.

Second, and the focus of this article, no matter how adequate our general theory of evolutionary change, we also yearn to document and understand the actual pathway of life's history. Theory, of course, is relevant to explaining the pathway (nothing about the pathway can be inconsistent with good theory, and theory can predict certain general aspects of life's geologic pattern). But the actual pathway is strongly *underdetermined* by our general theory of life's evolution. This point needs some belaboring as a central yet widely misunderstood aspect of the world's complexity. Webs and chains of historical events are so intricate, so imbued with random and chaotic elements, so unrepeatable in encompassing such a multitude of unique (and uniquely interacting) objects, that standard models of simple prediction and replication do not apply.

History can be explained, with satisfying rigor if evidence be adequate, after a sequence of events unfolds, but it cannot be predicted with any precision beforehand. Pierre-Simon Laplace, echoing the growing and confident determinism of the late 18th century, once said that he could specify all future states if he could know the position and motion of all particles in the cosmos at any moment, but the nature of universal complexity shatters this chimerical dream. History includes too much chaos, or extremely sensitive dependence on minute and unmeasurable differences in initial conditions, leading to massively divergent outcomes based on tiny and unknowable disparities in starting points. And history includes too much **contingency**, or shaping of present results by long chains of unpredictable antecedent states, rather than immediate determination by timeless laws of nature.

Life's History Is Not Predictable

Homo sapiens did not appear on the Earth, just a geologic second ago, because evolutionary theory predicts such an outcome based on themes of progress and increasing neural complexity. Humans arose, rather, as a fortuitous and contingent outcome of thousands of linked events, any one of which could have occurred differently and sent history on an alternative pathway that would not have led to consciousness. To cite just four among a multitude: (1) If our inconspicuous and fragile lineage had not been among the few survivors of the initial radiation of multicellular animal life in the **Cambrian** explosion 530 million years ago, then no **vertebrate** (backboned) animals would have inhabited the Earth at all. (Only one member of our **chordate** phylum, the genus *Pikaia*, has been found among these earliest fossils. This small and simple swimming creature, showing its allegiance to us by possessing a **notochord**, or dorsal stiffening rod, is among the rarest fossils of the Burgess Shale, our best preserved Cambrian fauna.) (2) If a small and unpromising group of **lobe-finned fishes** had not evolved fin bones with a strong central axis capable of bearing weight on land, then vertebrates might never have become terrestrial. (3) If a large extraterrestrial body had not struck the Earth 65 million years ago, then dinosaurs would still be dominant and mammals insignificant (the situation that had prevailed for 100 million years previously). (4) If a small lineage of primates had not evolved upright posture on the drying African savannas just two to four million years ago, then our ancestry might have ended in a line of apes that, like the chimpanzee and gorilla today, would have become ecologically marginal and probably doomed to extinction despite their remarkable behavioral complexity.

Therefore, to understand the events and generalities of life's pathway, we must go beyond principles of evolutionary theory to a paleontological examination of the contingent pattern of life's history on our planet—the single actualized version among millions of plausible alternatives that happened not to occur. Such a view of life's history is highly contrary both to conventional deterministic models of Western science and to the deepest social traditions and psychological hopes of Western culture for a history culminating in humans as life's highest expression and intended planetary steward.

Science can, and does, strive to grasp nature's factuality, but all science is socially embedded, and all scientists record prevailing "certainties," however hard they may be aiming for pure objectivity. Darwin himself, in the closing lines of *The Origin of Species*, expressed Victorian social preference more than nature's record in writing: "As natural selection works solely by and for the good of each being, all corporeal and mental endowments will tend to progress towards perfection."

Life's pathway certainly includes many features predictable from laws of nature, but these aspects are too broad and general to provide the "rightness" that we seek for validating evolution's particular results—roses, mushrooms, people, and so forth. Organisms adapt to, and are constrained by, physical principles. It is, for example, scarcely surprising, given laws of gravity, that the largest vertebrates in the sea (whales) exceed the heaviest animals on land (elephants today, dinosaurs in the past), which, in turn, are far bulkier than the largest vertebrate that ever flew (extinct **pterosaurs** of the **Mesozoic Era**).

Predictable ecological rules govern the structuring of communities by principles of energy flow and **thermodynamics** (more biomass in prey than in predators, for example). Evolutionary trends, once started, may have local predictability ("arms races," in which both predators and prey hone their defenses and weapons, for example a pattern that Geerat J. Vermeij of the University of California at Davis has called "escalation" and documented in increasing strength of both crab claws and shells of their gastropod prey through time). But laws of nature do not tell us why we have crabs and snails at all, why insects rule the multicellular world, why vertebrates rather than persistent microbial mats (stromatolites) exist as the most complex forms of life on the Earth.

Relative to the conventional view of life's history as an at least broadly predictable process of gradually advancing complexity through time, three features of the paleontological record stand out in opposition and shall therefore serve as organizing themes for the rest of this chapter: (1) The constancy of modal complexity throughout life's history; (2) the concentration of major events in short bursts interspersed with long periods of relative stability; and (3) the role of external impositions, primarily mass extinctions, in disrupting patterns of "normal" times. These three features, combined with more general themes of chaos and contingency, require a new framework for conceptualizing and drawing life's history, and this chapter therefore closes with suggestions for a different **iconography** of evolution.

Life's History Is Not Necessarily Progressive

The primary paleontological fact about life's beginnings points to predictability for the onset and very little for the particular pathways thereafter. The Earth is 4.5 billion years old, but the oldest rocks date to about 3.9 billion years because the Earth's surface became molten early in its history, a result of bombardment by large amounts of cosmic debris during the solar system's coalescence, and of heat generated by radioactive decay of short-lived **isotopes**. These oldest rocks are too **metamorphosed** by subsequent heat and pressure to preserve fossils (though some scientists interpret the proportions of carbon isotopes in these rocks as signs of organic production). The oldest rocks sufficiently unaltered to retain cellular fossils—African and Australian sediments dated to 3.5 billion years old do preserve **prokaryotic** cells (bacteria and **cyanobacteria**) and **stromatolites** (mats of sediment trapped and bound by these cells in shallow marine waters). Thus, life on the Earth evolved quickly and is as old as it could be. This fact alone seems to indicate an inevitability, or at least a predictability, for life's origin from the original chemical constituents of atmosphere and ocean.

No one can doubt that more complex creatures arose sequentially after this prokaryotic beginning—first **eukaryotic** (nucleated) cells, perhaps about two billion years ago, then multicellular animals about 600 million years ago, with a relay of highest complexity among animals passing from **invertebrates** (animals lacking backbones), to marine vertebrates and, finally (if we wish, albeit parochially, to honor neural architecture as a primary criterion), to reptiles, mammals, and humans. This is the conventional sequence represented in the old charts and texts as an "age of invertebrates," followed by an "age of fishes," "age of reptiles," "age of mammals," and "age of man" (to add the old gender bias to all the other prejudices implied by this sequence).

I do not deny the facts of the preceding paragraph but wish to argue that our conventional desire to view history as progressive, and to see humans as predictably dominant, has grossly distorted our interpretation of life's pathway by falsely placing in the center of things a relatively minor phenomenon that arises only as a side consequence of a physically constrained starting point. The most salient feature of life has been the stability of its bacterial mode from the beginning of the fossil record until today and, with little doubt, into all future time so long as the Earth endures. This is truly the "age of bacteria"—as it was in the beginning, is now and ever shall be.

For reasons related to the chemistry of life's origin and the physics of self-organization, the first living things arose at the lower limit of life's conceivable, preservable complexity. Call this lower limit the "left wall" for an architecture of complexity (see Figure 1.2). Since so little space exists between the left wall and life's initial bacterial mode in the fossil record, only one direction for future increment exists—toward greater complexity at the right. Thus, every once in a while, a more complex creature evolves and extends the range of life's diversity in the only available direction. In technical terms, the distribution of complexity becomes more strongly right skewed through these occasional additions

But the additions are rare and episodic. They do not even constitute an evolutionary series but form a motley sequence of distantly related taxa, usually depicted as eukaryotic cell, jellyfish, **trilobite**, nautiloid, **eurypterid** (a large relative of horseshoe crabs), fish, an amphibian, a dinosaur, a mammal, and a human being. This sequence cannot be construed as the major thrust or trend of life's history. Think rather of an occasional creature tumbling into the empty right region of complexity's space. Throughout this entire time, the bacterial mode has grown in height and remained constant in position. Bacteria represent the great success story of life's pathway. They occupy a wider domain of environments and span a broader range of biochemistries than any other group. They are adaptable, indestructible, and astoundingly diverse. We cannot even imagine how anthropogenic intervention might threaten their extinction, although we worry about our impact on nearly every other form of life. The number of *Escherichia coli* bacteria in the gut of each human being exceeds the number of humans that have ever lived on this planet!

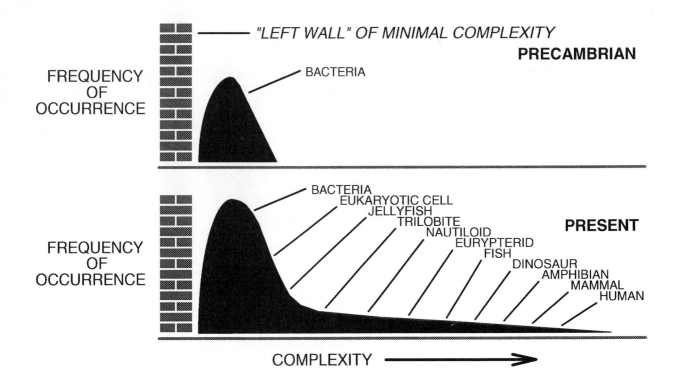

FIGURE 1.2 Progress does not rule (and is not even a primary thrust of) the evolutionary process. For reasons of chemistry and physics, life arises next to the "left wall" of its simplest conceivable and preservable complexity. This style of life (bacterial) has remained most common and most successful. A few creatures occasionally move to the right, thus extending the right tail of the distribution of complexity. Many always move to the left, but they are absorbed within space already occupied. Note that the bacterial mode has never changed in position, but just grown higher.

One might grant that complexification for life as a whole represents a pseudotrend based on constraint at the left wall but still hold that evolution within particular groups differentially favors complexity when the founding lineage begins far enough from the left wall to permit movement in both directions. Empirical tests of this interesting hypothesis are just beginning (as concern for the subject mounts among paleontologists), and we do not yet have enough cases to advance a generality. But the first two studies—by Daniel W. McShea of the University of Michigan on mammalian **vertebrae** (bony segments of the spinal column) and by George F. Boyajian of the University of Pennsylvania on **ammonite suture lines** (furrows at the junctions between chambers in fossil relatives of the chambered nautilus)—show no evolutionary tendencies to favor increased complexity.

Moreover, when we consider that for each mode of life involving greater complexity, there probably exists an equally advantageous style based on greater simplicity of form (as often found in parasites, for example), then preferential evolution toward complexity seems unlikely *a priori*. Our impression that life evolves toward greater complexity is probably only a bias inspired by parochial focus on ourselves, and consequent overattention to complexifying creatures, while we ignore just as many lineages adapting equally well by becoming simpler in form. The morphologically degenerate parasite, safe within its host, has just as much prospect for evolutionary success as its gorgeously elaborate relative coping with the slings and arrows of outrageous fortune in a tough external world.

RAPID EPISODIC BURSTS PUNCTUATE LONG PERIODS OF STABILITY

Even if complexity is only a drift away from a constraining left wall, we might view trends in this direction as more predictable and characteristic of life's pathway as a whole if increments of complexity accrued in a persistent and gradually accumulating manner through time. But nothing about life's history is more peculiar with respect to this common (and false) expectation than the actual pattern of extended stability and rapid episodic movement, as revealed by the fossil record.

Life remained almost exclusively unicellular for the first five-sixths of its history—from the first recorded fossils at 3.5 billion years to the first well documented multicellular animals less than 600 million years ago. (Some simple multicellular algae evolved more than a billion years ago, but these organisms belong to the plant kingdom and have no genealogical connection with animals.) This long period of unicellular life does include, to be sure, the vitally important transition from simple prokaryotic cells without **organelles** to eukaryotic cells with organellar nuclei, mitochondria, and other complexities of intracellular architecture but no recorded attainment of multicellular animal organization for a full three billion years. If complexity is such a good thing, and multicellularity represents its initial phase in our usual view, then life certainly took its time in making this crucial step. Such delays speak strongly against general progress as the major theme of life's history, even if they can be plausibly explained by lack of sufficient atmospheric oxygen for most of **Precambrian** time (which extends from the formation of the planet to the rise of multicellular animals at the onset of the immediately subsequent Cambrian Period) or by failure of unicellular life to achieve some structural threshold acting as a prerequisite to multicellularity.

More curiously, all major stages in organizing animal life's multicellular architecture then occurred in a short period beginning less than 600 million years ago and ending by about 530 million years ago—and the steps within this sequence are also discontinuous and episodic, not gradually accumulative. The first fauna, called **Ediacaran** to honor the Australian locality of its initial discovery but now known from rocks on all continents, consists of highly flattened fronds, sheets, and circlets composed of numerous slender segments quilted together. The nature of the Ediacaran Fauna is now a subject of intense discussion. These creatures do not seem to be simple precursors of later forms. They may constitute a separate and failed experiment in animal life, or they may represent a full range of **diploblastic** (two-layered) organization, of which the modern phylum **Cnidaria** (corals, jelly-fishes, and their allies) remains as a small and much altered remnant.

In any case, they apparently died out well before the Cambrian biota evolved. The Cambrian then began with an assemblage of bits and pieces, frustratingly difficult to interpret, called the "small shelly fauna." The subsequent main pulse, starting out 530 million years ago, constitutes the famous **Cambrian explosion** (see Figure 1.3) during which all but one modern phylum of animal life made a first appearance in the fossil record. (Geologists had previously allowed up to 40 million years for this event, but an elegant study, published in 1993, clearly restricts this period of phyletic flowering to a mere six million years.) The **Bryozoa**, a group of

FIGURE 1.3   Great diversity quickly evolved at the dawn of multicellular animal life during the Cambrian Period of Earth history (530 million years ago). The creatures shown here are all found in the Middle Cambrian Burgess Shale fauna of Canada. They include some familiar forms (sponges and brachiopods, known also as "lampshells") that have survived. But many creatures, such as the giant *Anomalocaris* (**41**)-the largest of all known Cambrian animals-did not live for long and are so anatomically peculiar (relative to surviors) that we cannot classify them among known phyla. Forms illustrated include (**1**) *Vauxia* (gracile form), (**2**) *Branchiocaris*, (**3**) *Opabinia*, (**4**) *Amiskwia*, (**5**) *Vauxia* (robust form), (**6**) *Molaria*, (**7**) *Aysheaia*, (**8**) *Sarotrocercus*, (**9**) *Nectocaris*, (**10**) *Pikaia*, (**11**) *Micromitra*, (**12**) *Echmatocrinus*, (**13**) *Chancelloria*, (**14**) *Pirania*, (**15**) *Choia*, (**16**) *Leptomitus*, (**17**) *Dinomischus*, (**18**) *Wiwaxia*, (**19**) *Naraoia*, (**20**) *Hyolithes*, (**21**) *Habelia*, (**22**) *Emeraldella*, (**23**) *Burgessia*, (**24**) *Leanchoilia*, (**25**) *Sanctacaris*, (**26**) *Ottoia*, (**27**) *Louisella*, (**28**) *Actaeus*, (**29**) *Yohoia*, (**30**) *Peronochaeta*, (**31**) *Selkirkia*, (**32**) *Ancalagon*, (**33**) *Burgessochaeta*, (**34**) *Sidneyia*, (**35**) *Odaraia*, (**36**) *Eiffelia*, (**37**) *Mackenzia*, (**38**) *Odontogriphus*, (**39**) *Hallucigenia*, (**40**) *Elrathia*, (**41**) *Anomalocaris*, (**42**) *Lingulella*, (**43**) *Scenella*, (**44**) *Canadaspis*, (**45**) *Marrella*, and (**46**) *Olenoides*.

sessile and colonial small marine organisms, do not arise until the beginning of the immediately subsequent Ordovician Period, but this apparent delay may be an artifact of failure to discover Cambrian representatives.

Although interesting and portentous events have occurred since—from the flowering of dinosaurs to the origin of human consciousness—we do not exaggerate greatly in stating that the subsequent history of animal life amounts to little more than variations on anatomical themes established during the Cambrian explosion within a scant six million years. Three billion years of unicellularity, followed by six million years of intense creativity and then capped by more than 500 million years of variation on set anatomical themes can scarcely be read as a predictable, inexorable, or continuous trend toward progress or increasing complexity!

We do not know why the Cambrian explosion could establish all major anatomical designs so quickly. An "external" explanation based on ecology seems attractive; the Cambrian explosion represents an initial filling of the "ecological barrel" of niches for multicellular organisms, and any experiment found a space.

The barrel has never emptied since—even the great mass extinctions left a few species in each principal role, and their occupation of ecological space forecloses opportunity for fundamental novelties. But an "internal" explanation based on genetics and development also seems necessary as a complement: The earliest multicellular animals may have maintained a flexibility for genetic change and embryological transformation that became greatly reduced as organisms "locked in" a set of stable and successful designs.

In any case, this initial period of both internal and external flexibility yielded a range of anatomies among invertebrates (see Figure 1.4) that may have exceeded—in just a few million years of production—the full scope of animal form in all the Earth's environments today (after more than 500 million years of additional time for further expansion). Scientists are divided on this question.

Some claim that the anatomical range of this initial explosion exceeded that of modern life, as many early experiments died out and no new phyla have ever arisen. But scientists most strongly opposed to this view allow that Cambrian diversity at least equaled the modern range, so even the most cautious opinion holds that 500 million subsequent years of opportunity have not expanded the Cambrian range, achieved in just six million years. The Cambrian explosion was the most remarkable and puzzling event in the history of life.

Moreover, we do not know why most of the early experiments died, while a few survived to become our modern phyla. It is tempting to say that the victors won by virtue of greater anatomical complexity, better ecological fit, or some other predictable feature of conventional **Darwinian struggle**. But no recognized traits unite the victors, and the radical alternative must be entertained that each early experiment received little more than the equivalent of a ticket in the largest lottery ever played out on our planet—and that each surviving lineage, including our own phylum of vertebrates, inhabits the Earth today more by the luck of the draw than by any predictable struggle for existence. The history of multicellular animal life may be more a story of great reduction in initial possibilities, with stabilization of lucky survivors, than a conventional tale of steady ecological expansion and morphological progress in complexity.

Finally, this pattern of long **stasis** (relative quiescence and stability), with change concentrated in rapid episodes that establish new equilibria and stable biotic relationships, may be quite general at several scales of time and magnitude, forming a kind of **fractal** pattern in self-similarity. According to the punctuated equilibrium model of speciation (discussed in Chapter 3), trends within lineages occur by accumulated episodes of geologically instantaneous speciation, rather than by gradual change within continuous populations (like climbing a staircase rather than rolling a ball up an inclined plane).

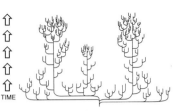

FIGURE 1.4 New iconography of life's tree shows that maximal diversity in anatomical forms (not in number of species) is reached very early in life's multicellular history. Later times feature extinction of most of these initial experiments and enormous success within surviving lines. This success is measured in the proliferation of species but not in the development of new anatomies. Today we have more species than ever before, although they are restricted to fewer basic anatomies.

MASS EXTINCTIONS DISRUPT PATTERNS OF "NORMAL" TIMES

Even if evolutionary theory implied a potential internal direction for life's pathway (although previous facts and arguments in this essay cast doubt on such a claim), the occasional imposition of a rapid and substantial, perhaps even truly catastrophic, change in environment would have intervened to stymie the pattern. These environmental changes trigger mass extinction of a high percentage of the Earth's species and may so derail any internal direction and so reset the pathway that the net pattern of life's history looks more capricious and concentrated in episodes than steady and directional. Mass extinctions have been recognized since the dawn of paleontology; the major divisions of the geologic time scale were established at boundaries marked by such events. But until the revival of interest that began in the late 1970s, most paleontologists treated mass extinctions only as intensifications of ordinary events, leading (at most) to a speeding up of tendencies that pervaded normal times. In this gradualistic theory of mass extinction, these events really took a few million years to unfold (with the appearance of suddenness interpreted as an artifact of an imperfect fossil record), and they only made the ordinary occur faster (more intense Darwinian competition in tough times, for example, leading to even more efficient replacement of less adapted by superior forms).

The reinterpretation of mass extinctions as central to life's pathway and radically different in effect began with the presentation of data by Luis and Walter Alvarez in 1979, indicating that the impact of a large extraterrestrial object (they suggested an asteroid seven to 10 km in diameter) set off the last great extinction that marks the boundary between the **Cretaceous** and **Tertiary** Periods of Earth history 65 million years ago. Although the Alvarez hypothesis initially received very skeptical treatment from scientists (a proper approach to highly unconventional explanations), the case now seems virtually proved by discovery of the "smoking gun," a crater of appropriate size and age located off the Yucatan peninsula in Mexico.

This reawakening of interest also inspired paleontologists to tabulate the data of mass extinction more rigorously. Work by David M. Raup, J. John Sepkoski, Jr., and David Jablonski of the University of Chicago has established that multicellular animal life experienced five major (end of **Ordovician**, late **Devonian**, end of **Permian**, end of **Triassic**, end of Cretaceous) and many minor mass extinctions during its 530-million-year history. We have no clear evidence that any but the last of these events was triggered by catastrophic impact, but such careful study leads to the general conclusion that mass extinctions were more frequent, more rapid, more extensive in magnitude, and more different in effect than paleontologists had previously realized. These four properties encompass the radical implications of mass extinction for understanding life's pathway as more contingent and chancy than predictable and directional.

Mass extinctions are not random in their impact on life. Some lineages succumb and others survive as sensible outcomes based on presence or absence of evolved features. But especially if the triggering cause of extinction be sudden and catastrophic, the reasons for life or death may be random with respect to the original value of key features when first evolved in Darwinian struggles of normal times. This "different rules" model of mass extinction imparts a quirky and unpredictable character to life's pathway based on the evident claim that lineages cannot anticipate future contingencies of such magnitude and different operation.

To cite two examples from the impact-triggered Cretaceous-Tertiary extinction 65 million years ago: First, an important study published in 1986 noted that **diatoms** survived the extinction far better than other single-celled plankton (primarily **coccoliths** and **radiolaria**). This study found that many diatoms had evolved a strategy of dormancy by encystment, perhaps to survive through seasonal periods of unfavorable conditions (months of darkness in polar species as otherwise fatal to these photosynthesizing cells; sporadic availability of silica needed to construct their skeletons). Other planktonic cells had not evolved any mechanisms for dormancy. If the terminal Cretaceous impact produced a dust cloud that blocked light for several months or longer (one popular idea for a "killing scenario" in the extinction), then diatoms may have survived as a fortuitous result of dormancy mechanisms evolved for the entirely different function of weathering seasonal droughts in ordinary times. Diatoms are not superior to radiolaria or other plankton that succumbed in far greater numbers; they were simply fortunate to possess a favorable feature, evolved for other reasons, that fostered passage through the impact and its sequelae.

Second, we all know that dinosaurs perished in the end Cretaceous event and that mammals therefore rule the vertebrate world today. Most people assume that mammals prevailed in these tough times for some reason of general superiority over dinosaurs. But such a conclusion seems most unlikely. Mammals and dinosaurs had coexisted for 100 million years, and mammals had remained rat-sized or smaller, making no evolutionary "move" to oust dinosaurs. No good argument for mammalian prevalence by general superiority has ever been advanced, and fortuity seems far more likely. As one plausible argument, mammals may have survived partly as a result of their small size (with much larger, and therefore extinction resistant, populations as a consequence, and less ecological specialization with more places to hide, so to speak). Small size may not have been a positive mammalian adaptation at all, but more a sign of inability ever to penetrate the dominant domain of dinosaurs. Yet this "negative" feature of normal times may be the key reason for mammalian survival and a prerequisite to my writing and your reading this chapter today.

FREUD'S PEDESTAL AND THE DARWINIAN REVOLUTION

Sigmund Freud often remarked that great revolutions in the history of science have but one common, and ironic, feature: They knock human arrogance off one pedestal after another of our previous conviction about our own self-importance. In Freud's three examples, Copernicus moved our home from center to periphery; Darwin then relegated us to "descent from an animal world"; and, finally (in one of the least modest statements of intellectual history), Freud himself discovered the unconscious and exploded the myth of a fully rational mind.

In this wise and crucial sense, the Darwinian revolution remains woefully incomplete because, even though thinking humanity accepts the fact of evolution, most of us are still unwilling to abandon the comforting view that evolution means (or at least embodies a central principle of) progress defined to render the appearance of something like human consciousness either virtually inevitable or at least predictable. The pedestal is not smashed until we abandon progress or complexification as a central principle and come to entertain the strong possibility that *Homo sapiens* is but a tiny, late-arising twig on life's enormously arborescent bush—a small bud that would almost surely not appear a second time if we could replant the bush from seed and let it grow again.

Primates are visual animals, and the pictures we draw betray our deepest convictions and display our current conceptual limitations. Artists have always painted the history of fossil life as a sequence from invertebrates, to fishes, to early terrestrial amphibians and reptiles, to dinosaurs, to mammals and, finally, to humans. There are no exceptions; all sequences painted since the inception of this genre in the 1850s follow the convention (see Figure 1.5).

Yet we never stop to recognize the almost absurd biases coded into this universal mode. No scene ever shows another invertebrate after fishes evolved, but invertebrates did not go away or stop evolving! After terrestrial reptiles emerge, no subsequent scene ever shows a fish (later oceanic tableaux depict only such returning reptiles as **ichthyosaurs** and **plesiosaurs**). But fishes did not stop evolving after one small lineage managed to invade the land. In fact, the major event in the evolution of fishes, the origin and rise to dominance of the **teleosts**, or modern bony

FIGURE 1.5 Classical representations of life's history reveal the biases of viewing evolution as progress and complexification. These paintings by Charles R. Knight from a 1942 issue of *National Geographic*, first show invertebrates of the Burgess Shale (A), but once fishes evolve (B), no subsequent scene ever shows another invertebrate although they did not go away or stop evolving. When land vertebrate lineages arise (C), we never see another fish, even though return of land vertebrate lineages to the sea may be depicted (D). The sequence always ends with mammals (E)—even though fishes, invertebrates, and reptiles are still thriving—and, of course, humans (F).

fishes, occurred during the time of the dinosaurs and is therefore never shown at all in any of these sequences—even though teleosts include more than half of all species of vertebrates! Why should humans appear at the end of all sequences? Our group (formally, a taxonomic **order**) of primates is ancient among mammals, and

FIGURE 1.5 *Continued*

many other successful lineages arose later than we did.

We will not smash Freud's pedestal and complete Darwin's revolution until we find, grasp, and accept another way of drawing life's history. J. B. S. Haldane proclaimed nature "queerer than we can suppose," but these limits may only be socially imposed conceptual locks rather then inherent restrictions of our neurology. New icons might break the locks. Trees—or rather copiously and luxuriantly branching bushes—rather than ladders and sequences hold the key to this conceptual transition.

We must learn to depict the full range of variation, not just our parochial perception of the tiny right tail of most complex creatures. We must recognize that this tree may have contained a maximal number of branches near the beginning of multicellular life and that subsequent history is for the most part a process of elimination and lucky survivorship of a few, rather than continuous flowering, progress, and expansion of a growing multitude. We must understand that little twigs are contingent nubbins, not predictable goals of the massive bush beneath. We must remember the greatest of all Biblical statements about wisdom: "She is a tree of life to them that lay hold upon her; and happy is every one that retaineth her."

Further Reading

Gould, S. J. 1987. *Wonderful Life: The Burgess Shale and the Nature of History* (New York: Norton), 347 pp.

Gould, S. J. (Editor). 1993. *The Book of Life* (New York: Norton), 256 pp.

Stanley, S. M. 1987. *Extinction: A Scientific American Book* (New York: W. H. Freeman), 242 pp.

CHAPTER 2

DATING THE TIMELINE OF LIFE'S HISTORY

JON P. DAVIDSON[1]

INTRODUCTION

Evolution is defined as a process of change through time. The evolution of life on Earth, formalized in Darwin's theory, is inextricably linked to the evolution of the atmosphere and oceans as well as to changes of the planet's surface, driven ultimately by geologic forces and climate. Direct evidence of biologic evolution, albeit very imperfect, is preserved as shells, burrows, bones, and impressions—structures we know as fossils. But just as we need to order the pages of a book to make sense of the story it tells, we must be able to determine the order of fossils through time to understand evolution. This chapter describes the ways by which this ordering has been accomplished to yield a record of biologic history that is now well-calibrated through geologic time.

MEASURING TIME—LOOKING FOR A SENSE OF DIRECTION

It may not seem much of a problem to determine the direction time's arrow flies. We of course grow older, rather than younger, and must pay the price by putting up with a great number of obvious physiological changes. And the pace of technological change enables us to have a keen sense of the age of items in our surroundings—we can distinguish a movie or a television show that is 10 years old from one that is 25 simply from the automobiles or clothes of the period.

On the other hand, many systems in nature do not carry such obvious hints of the direction of time's arrow but instead change in a cyclic, repetitive fashion. Indeed, in the geosciences the concept of cycles aids greatly in understanding many different systems. The **hydrologic cycle**, for example, describes the flux of water at Earth's surface—it's evaporation from the oceans, condensation as rain or snow, and ultimate recycling back to the oceans by streams, rivers, glaciers, or even the slow movement of subterranean groundwaters. The actual path and time taken to return to the oceans varies for each water molecule. Some move rapidly through the system whereas others reside temporarily in standing bodies of water such as lakes,

[1]Department of Earth and Space Sciences, University of California, Los Angeles, CA 90095.

and still others pass through organisms where they play a role in living processes. But we would not notice much difference in "snapshots" of the hydrologic system taken at geologically different times—there would be no easy way to detect time's arrow, to distinguish an ancient snapshot from one more recent because both would show the same cyclic system with all parts in operation as they always have been, for billions of years, ever since a sizable reservoir of water was present on the planet's surface.

Other cycles in nature are related to the fundamental cyclicity of the planets, and indeed the solar system. Because the Earth spins about its axis we experience day-night cycles that affect the growth of organisms, most notably, of course, plantlife. And because the Earth is tilted on its axis and cycles the Sun we experience the seasons that define one of our most important measures of time, the year. The orbital dynamics of the Earth relative to the Sun also give rise to cyclic variations that can affect long-term climate. Changes in the tilt (the **precession**) of Earth's axis and the deviation of its orbit around the Sun from a circular path (the orbit's **eccentricity**) can alter the effective distance of Earth's poles from the Sun. These periodic changes are called **Milankovitch cycles** and appear to play a role in bringing on ice ages, if not by actually triggering them, then at least by influencing the advance or retreat of ice sheets as ice volumes grow or shrink in polar regions.

Geological processes also are commonly cyclic. For example, because the processes of rock formation followed by erosion then formation of new rocks is cyclic, we can relate different types of rocks through what is known as the **rock cycle** (Figure 2.1). Mountain belts are crumpled uplifted parts of the Earth's crust formed by **tectonic** (crust-deforming) processes where huge thick masses of rock (**crustal plates**) slowly but inexorably over time grind against one another. During

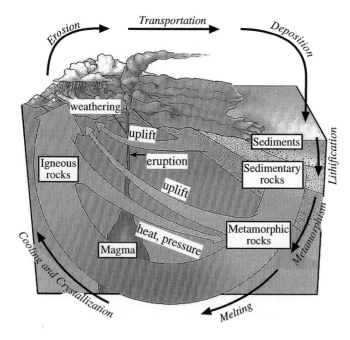

FIGURE 2.1 The Rock Cycle showing the way geologists conceptualize the relations between Earth materials (boxes) and the processes by which they are related (arrows). In the top part of the diagram, the hydrologic cycle works over short time scales to weather and erode continents and deposit sediments in the oceans. The cyclic processes shown have operated since early in Earth's history. (Modified after Fig. 4.29 in Davidson, Reed and Davis, 1997; used with permission.)

these events some parts of the crust are pressure-cooked to form physically and geochemically altered (**metamorphic**) rocks, whereas others are carried to depth and completely melted (masses that on cooling crystallize to form what are known as **igneous** rocks). Over time, of course, the metamorphic and igneous rocks that make up mountains are worn down by wind and rain to produce rock and mineral particles that are carried by streams and rivers to the oceans where they are sedimented in layers that eventually solidify to form **sedimentary** rocks. But since the new-formed sediments are deposited in the vicinity of the same plate boundary that produced the mountain belt in the first place, they are destined to again be cycled through the system, from sedimentary to metamorphic and igneous rocks then back to sediments, over and over.

The rock cycle is the foundation of one of the most useful concepts in all of the geosciences, the **Principle of Uniformitarianism** set out by James Hutton in the late 18th century (Hutton, 1788). The principle of uniformitarianism is often paraphrased as "the present is the key to the past," the idea being that because natural processes operating today have *always* operated we can use them as a guide in understanding events throughout all geologic time. The enormous duration of Earth history (arguably Hutton's greatest insight) makes the cyclicity of geologic processes especially striking and led Hutton to conclude that the rock record has "no vestige of a beginning, no prospect of an end." Taken to an extreme the concept of uniformitariansim can be applied both to the processes themselves and to the magnitudes and rates at which they occur, a philosophy known as **gradualism** and championed by the British geologist Sir Charles Lyell (Lyell, 1877).

The concept of time as having both a "straight line" and "cyclic" nature is discussed eloquently by Stephen Jay Gould in *Time's Arrow, Time's Cycle* (Gould, 1987). Time, of course, always moves in only one direction, but it's a question of one's perception whether it can also be cyclic (as captured, for example, in the common phrase "what goes around comes around").

Perhaps the simplest way to understand the nature of time is by considering the **Second Law of Thermodynamics**, which states that for any spontaneous process in a system, **entropy**—the amount of "disorder" in the system—always increases. Entropy can never decrease (although by some processes is only little changed). For instance, if a cup of coffee is knocked to the floor it will shatter, markedly increasing the entropy ("disorder") of the system, and if one were to make a video recording of the event and play it back it would be easy to tell whether the film was running forward or backward. In this and most situations the direction of time's arrow is obvious—although Steven Hawking (1988) points out that disorder (entropy) increases with time because we of course measure time in that direction!

EVOLUTION VERSUS REVOLUTION

In geology, perhaps the strongest sense of time's directionality is revealed by biological evolution. Unlike many other aspects of the geosciences, the fossil record is additive rather than cyclic and this provides a strong sense of directionality. Life evolved from simple to more complex, starting with single cells and over time adding increasingly complicated many-celled organisms, a continuum of life forms that enables us to calibrate the rock record using the fossils it contains. Indeed, by comparing the similarity of fossils found in rocks of various ages to organisms living today, progressive changes in the fossil record were recognized even before Darwin proposed the theory of evolution and were used to define what then were known as the "Primary," "Secondary," and "Tertiary" Periods of the geologic record (of which only the Tertiary remains in use).

It is not surprising, therefore, that fossils can be used to sort out the timing of various phases in cyclic geologic processes. For example, although mountain belts have formed, eroded, and reformed throughout Earth's history, snapshots of their rocks that included lifeforms of the time would permit us to figure out which belts were younger, which more ancient.

LOOKING FOR A TIME SCALE: METHODS OF DATING

Hutton, Lyell, and Darwin all were convinced the Earth was ancient, but just how ancient no one knew. And to Darwin and his theory the question was crucial—for "survival of the fittest, descent with modification" to work, it needed time, and lots of it.

How old is the Earth? The answer sets a limit to the measuring stick for dating events in Earth history (Figure 2.2). In the Western world, early scholars attempted to establish the age of the Earth by making genealogical calculations based on the Biblical lineage beginning with Adam and Eve. Almost all arrived at a similar and very young age for the Earth (Figure 2.2), the most often quoted estimate that of Archbishop Usher who in 1654 calculated that the Earth was created on October 23rd of the year 4004 BC.

But scientists in the 17th and 18th centuries surmised from the rates at which processes were observed in nature that the Earth must be much older. Hutton's

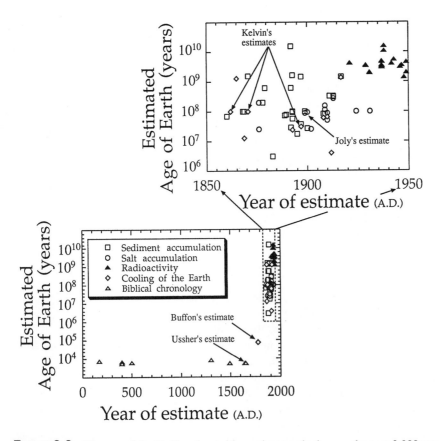

FIGURE 2.2 The age of the Earth estimated by various methods over the past 2,000 years (based on data from Dalrymple, 1991).

astute observations of rates of erosion and deposition led him to conclude that vast amounts of time would be needed to account for many features obvious in rock outcroppings. For instance, from Siccar Point on the southeast coast of Scotland he described a particularly fine example of what is called an **unconformity** (Figure 2.3). He reasoned that the lower series of rocks was deposited over a long period of time and then uplifted and deformed in a mountain-building episode that accounts for their present-day tilt. Erosion then exposed the surface made up of tilted rocks which came to be submerged and on top of which the upper sequence of sediments was deposited to produce the unconformity. Subsequent uplift exposed the strata as they are today.

These various events and processes would have taken a long, long time. Obviously, deposition of the sediments and their solidification to sedimentary rocks could happen only over a very long period if the rates were even close to those Hutton observed in rivers and oceans, but the time represented by the gap in the rock record—represented by the unconformity when the lower part of the sequence was uplifted, tilted, and eroded—may have been even longer. All told, the time required would have been far greater than that calculated by religious scholars, but there still was no way to measure how long the time may have been.

Relative and Absolute Dating

There are two different ways of talking about age, "relative" (younger as opposed to older) and "absolute" (the chronological age, given in years).

FIGURE 2.3 Geology students inspecting "Hutton's unconformity," a prominent angular discordance between two sets of rock layers exposed at Siccar Point, Scotland. Lower strata of the succession (the beds in the foreground) were deposited as horizontal layers then tilted and uplifted to their present near-vertical orientation. Subsequent erosion created a surface on which the later (upper) layers of sediment (the slightly tilted beds at the left) were deposited. After deposition these also were uplifted and slightly tilted to their present position where both they and the lower strata are now subject to erosion. James Hutton recognized that these processes could have happened only over an enormous span of time.

In a population of trees in a forest, for example, one might argue that the larger trees are older than the smaller ones because they've grown for a longer time. The trees' height gives a sense of their *relative* age but does not tell whether they are tens of years old or perhaps much older, dating from hundreds of years ago. If we knew their rates of growth we perhaps could calculate their ages, but growth rates vary even among individuals of a single species and are affected by changes in climate or events such as severe wind storms or forest fires.

But if we were to cut through the trunks of the trees and count their tree rings we could figure out their *absolute* age, in years. Each ring represents a yearly growth period in which the large thin-walled cells formed in the spring are easy to distinguish from the smaller thick-walled cells laid down in the autumn before growth ceases with the onset of winter. A similar principle can be used to identify yearly layers known as **varves** in the sedimentary deposits of glacial lakes—dark horizons rich in organic matter are laid down in the winter when the lakes are ice-covered and almost no mineral debris is carried into the lakes, whereas lighter colored layers rich in mineral detritus form during the spring and summer when abundant sediment is derived from glacial meltwaters.

Relative ages of rocks, fossils, and geologic events can be determined by a number of methods which I will soon discuss. But *absolute* ages are required to develop an accurate perspective on geologic history and to address the important question of rates of evolution, both of geologic processes and of life.

Stratigraphy

One of the simplest methods of relative dating is by use of stratigraphy and the stratigraphic relations among rock units. Because of the influence of gravity, sedimentary strata are deposited on the Earth's surface such that in a series of more or less horizontal rock beds those at the bottom are oldest while those at the top are youngest—a simple set of relations known as the **Law of Superposition** and first formalized in the 17th century by an Italian geologist named Nicholas Steno. Thus, in a succession of flat-lying beds such as encountered in the Grand Canyon of Arizona (Figure 2.4A), the rocks that form the lip of the canyon are youngest whereas those deep within the canyon are older. The Law of Superposition can tell us the relative ages of rock units in a stacked series but not their actual, absolute ages. And there are yet other limitations on the usefulness of this simple principle (Figure 2.4B):

1. Strata are limited in their lateral spread—they can be thick in one area but then pinch out to nothing at some distance away. If a single bed is traced laterally it will either thin and disappear (reflecting the margin of the depositional basin or depression in which it was deposited), merge into another type of rock (formally, a "**facies** variation" which reflects changes in the depositional setting), or end abruptly, cut off by a geologic structure such as an earthquake **fault** where the rocks have been shifted by massive movements of the crust. Thus, stratigraphic relations established in one region often cannot easily be used to tell the relative ages of rock layers someplace else.

2. Steno's Law of Superposition cannot be applied to strongly deformed strata since their original horizontal orientation is needed to establish the order of deposition.

3. The Law of Superposition is useful mostly to tell the relative ages of sedimentary, rather than metamorphic or igneous rocks (although widespread lava flows, igneous rocks, can also form thick sequences of horizontal layers). And

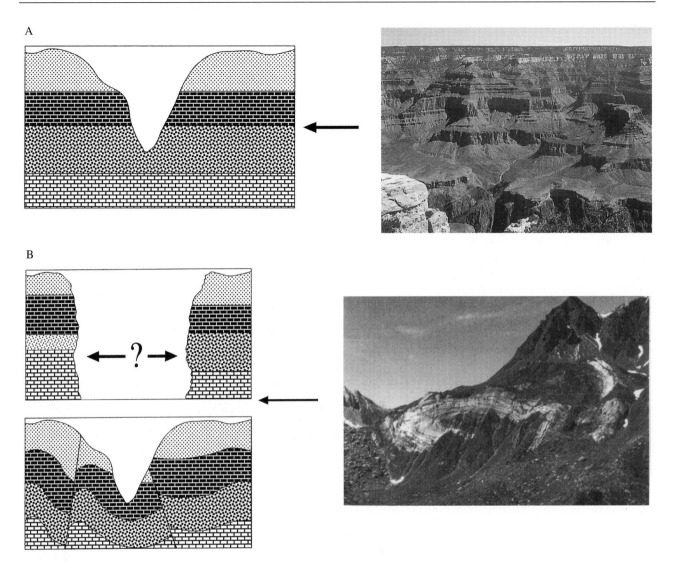

FIGURE 2.4   (A) According to the principles of stratigraphy and the Law of Superposition, individual rock units can be laterally correlated (shown to represent the same time), with the strata being progressively younger from bottom to top of a rock sequence such as this. For an extensive flat-lying sequence of rocks, such as that shown here in the Grand Canyon, Arizona, correlations can be made quite easily. (B) But correlation can be difficult if a sequence is subject to extensive erosion between the deposition of sequences of rock or if there are lateral changes in the character of the rock layers. Folding, faulting, and other deformation of the strata may add to these problems, making it difficult to decipher the original order in which the sequence was deposited.

although sedimentary rocks cover some 75% of the surface of continents they make up only about 5% of their volumes and are consequently far less abundant than rocks to which the law is difficult to apply.

The technique of tracing a rock layer based on its characteristics (such as color, grain size, mineral makeup, and so forth) is known as **lithostratigraphy**. But it is important to realize that the layers of rock identified by lithostratigraphy do not necessarily represent "slices of time." Even in a single basin the conditions that govern sedimentation can shift location from one time to another so that the same type of rock can be laid down at different places at different times. For this reason, surfaces between rock beds commonly cut *across* time horizons. By analogy, a Highway Department truck salting a long stretch of icy road leaves as a record of its progress a single layer of salt spread on the road, but the layer is of course not deposited at a single instant of time. Just as the same rock bed can be older at one place and younger someplace else, the salt layer deposited on the stretch of road where the truck began work is older than the same layer left on the road where it finished.

Accumulation Rates

Some of the earliest attempts to actually measure geologic time, rather than determine only the relative ages of geologic units, centered on the use of accumulation rates—particularly of sediments and of the buildup of salt in the world's oceans. These were a direct application of Hutton's principle that the present is the key to the past.

For sedimentary rocks, the idea was that if the rate at which sediment accumulates today were known and the thickness of a sedimentary unit measured, then the time for the rock to form might be accurately estimated. This approach turned out to have many pitfalls—sedimentation rates vary widely place to place and time to time; sedimentary processes commonly are episodic rather than continuous and uniform; and the gaps of time when sediments are not deposited in a sedimentary succession are difficult to determine. Nevertheless, between 1860 and the early 1900s estimates on the basis of sediment accumulation studies showed that the Earth must be very much older than suggested by Biblical chronologies (Figure 2.2).

An even more ingenious attempt to estimate the age of the Earth was based on calculating the time required to leach salts from exposed land masses and carry the salts in streams and rivers to the world's oceans where they were thought to have accumulated since the beginnings of the planet. Although not the first to propose this idea, Irish geologist John Joly is generally credited with its most refined usage.

To use this method it had to be assumed that the oceans started out as freshwater and that modern rates of weathering, erosion, transport of salts, and so forth are representative of the geologic past (in essence, a uniformitarian perspective). These assumptions were at least partly incorrect, but an even more telling flaw was that the method was formulated at the turn of the last century when little was known about the ocean floor and nothing about plate tectonics so that salt added by interaction of seawater with cooling lavas at mid-ocean ridges was not accounted for. Although the method proved wrong because certain of its assumptions were invalid it, too, suggested the Earth is much older than was previously thought (Figure 2.2).

Cooling Rates

A variation of the use of accumulation rates, based on the rate of cooling of a hot primordial Earth to its present temperature, was also used to estimate the age of the planet. The simple assumption was that the Earth began molten and subsequently cooled, so if the rate of this cooling were known it could then be estimated over what period it happened. In the late 18th century, French natural scientist G. L. de Buffon estimated cooling rates from a series of experiments using different-sized spheres of steel and concluded that some 75,000 years would have been needed for the Earth to cool from a molten state. While once again the calculations were based on invalid assumptions, Buffon's work provided one of the earliest scientific arguments countering the widely accepted Bible-based estimates.

A more sophisticated attempt was developed in 1846 by British physicist Lord Kelvin who derived an age for the Earth of 20 to 30 million years. By Kelvin's time, the mid-1800s, a great deal of geological study had already been carried out, building on the work of Hutton and Lyell and coming to the inescapable conclusion that the Earth was indeed "ancient." In fact, Kelvin's estimate of 20 to 30 million years seemed too short to account for the very thick rock record that by then was already documented. Although Kelvin, a physicist, claimed his estimates were founded on rigorous numerical physics and thus superior to the qualitative arguments of geologists, he too based his calculations on what have since proven incorrect assumptions. Most importantly, he assumed the Earth lacks an internal source of heat, but half a century later the discovery of radioactivity led to realization that radioactive heat would make the Earth's cooling very much slower than Kelvin imagined (a breakthrough that would pave the way to development of a tool, isotopic dating, to calibrate the geologic time-scale with absolute ages).

Fossil Correlation

Fossils have been known for thousands of years, but even up to the late 1700s were widely interpreted as lithified remains of plants and animals that found their way into crevices in rocks where they are now preserved, or remnants of organisms drowned in the Biblical Great Flood (see Chapter 5). The true significance of fossils was recognized only in the 18th century when detailed studies of geology (including preparation of the earliest geological maps) were first carried out.

In England, during the late 18th century, surveyor William Smith constructed geological maps showing the spread of strata he had repeatedly seen in mines and canal excavations. Although there were covered areas between the exposures, he was able to identify the same beds extending from one place to another by using the distinctive fossils they contained (Figure 2.5). At about the same time, French scientists Georges Cuvier and Alexandre Brogniart published geological maps of northern France, with Cuvier noting:

> These fossils are generally the same in corresponding beds, and present tolerably marked differences of species from one group of beds to another. It is a method of recognition which up to the present has never deceived us.

The fossils to which Cuvier referred are known now as **index fossils**. Not all fossils can be used to show that rock beds in one place are the same age as those someplace else (that is, to **correlate** the strata). Some fossil species change little over long periods of time (and so are not good for identification of short time periods), whereas others are only found in uncommon environments (and so are not spread widely enough to be used). To be useful as an index fossil a species must be easily recognizable, exist for only a relatively short period of time, and be geographically widespread.

Once it was recognized that fossils could be used to establish the relative ages of sediments in which they are entombed they could then be combined with stratigraphy to establish the relative ages of rock units over wide areas. The development of a geologic time-scale for the entire world was based on this principle, but until the methods of radioactive dating eventually came to be developed the scale could not be calibrated in terms of thousands, millions, or billions of years.

It is important to note that although fossil correlation (known also as **biostratigraphy**) works because of biological evolution, by itself it is not a proof of evolution. The fossils are simply used as tracers, "labels," to identify strata of the same age. By analogy, we could use the features of everyday garbage—beer cans, soda bottles, and the like—to date the layers of a landfill. Their shapes, colors, and labeling would allow us to date a particular layer, often to within a year or so, but such man-made artifacts are of course not products of biological evolution.

Radioactive Dating

In 1896 French physicist Henri Bequerel discovered radioactivity, a breakthrough that eventually came to be the basis by which to determine the absolute ages of rocks and thereby calibrate all of Earth history.

   Radioactivity is the spontaneous change, "decay," of one isotope (known as the parent isotope) of a chemical element to another (the daughter). Most elements come in more than one naturally occurring isotopic form. The nuclear cores of the atoms of all isotopic forms of any element share the same number of positively charged protons (a number which identifies the element) but for different isotopes have different numbers of neutrons and therefore different masses. Unlike most isotopes which are stable, completely immutable, those that decay through radioactivity have nuclei in which the numbers of neutrons and protons can spontaneously change to a more energetically favorable composition.

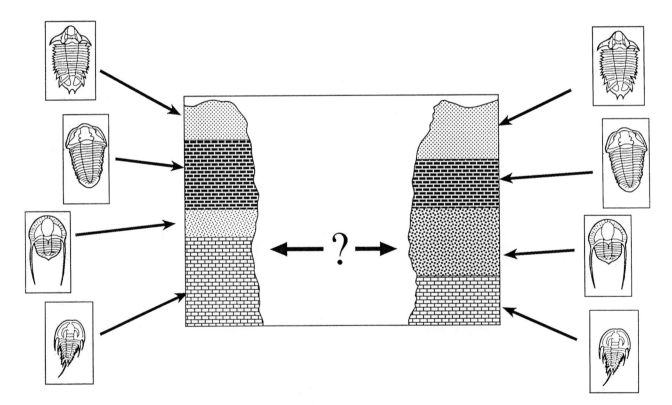

FIGURE 2.5 In biostratigraphy, distinctive index fossils are used to label rock layers enabling them to be correlated as time equivalents even if they are deformed or separated by large distances.

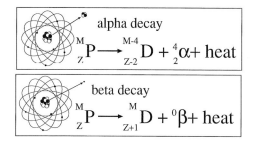

FIGURE 2.6 The two main types of radioactive decay. *Alpha* (α) *decay* happens when two neutrons and two protons (an alpha particle) are ejected from an atomic nucleus, whereas *beta* (β) *decay* is the ejection of an electron (beta particle) from a neutron, which consequently becomes a proton. Electron capture (not shown) is essentially the opposite of beta decay—an electron from the inner shell an atom falls into the nucleus and combines with a proton to form a neutron. The decay from parent (P) to daughter (D) involves changes in atomic number (Z) and/or mass (M).

The decay takes two main forms (Figure 2.6): Alpha decay occurs when an alpha (α) particle (two neutrons and two protons) is ejected from a nucleus. Compared to the parent isotope, the resulting daughter has an atomic mass that is less by four mass units and an atomic number less by two. Beta decay occurs when a neutron transforms to a proton, ejecting a beta (β) particle (an electron). In this case the atomic mass is unchanged and the daughter has an atomic number greater by one unit than the parent.

In both alpha and beta decay, heat is produced which because of the abundance of radioactive elements within the Earth heats the interior of the planet. Although this fact invalidated Lord Kelvin's estimates for the age of the Earth (Figure 2.2), studies beginning in the early 1900s showed that radioactive decay has a clocklike behavior that can serve to date geologic events.

For any radioactive isotope, the rate at which the parent decays is governed by what is known as the decay constant, lambda, "λ." A more familiar expression of the decay rate is the **half-life** (abbreviated "$t_{1/2}$"), the time it takes for one-half of the original number of parent atoms to decay to daughter atoms. At any time, the number of atoms of the parent isotope (N) that remain unchanged to daughter atoms is governed both by the number of parent atoms originally present and the decay constant of the radioactive isotope, a relationship expressed mathematically as:

$$dN/dt = -\lambda N.$$

The solution to this simple differential equation is:

$$N = N_{0}e^{-\lambda t}$$

where "N_0" is the number of atoms of the parent isotope present at the beginning of the process, time zero (t = 0).

How this decay changes over time (expressed as half-lives) can be seen in Figure 2.7 which shows that the number of parent atoms decreases as daughters are radioactively produced. From this it is clear that the parent to daughter ratio varies through time, so that if we know the decay constant of a radioactive isotope we can then use the parent-daughter ratio to measure the age of a sample containing that isotope.

Reality is not quite so simple. In samples from nature we do not know how much of the radioactive parent was present at t = 0. This amount (N_0), however, can

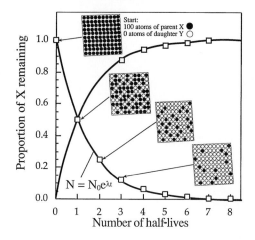

FIGURE 2.7 In radioactive decay, if parent isotope "X" decays to daughter isotope "Y," then after a time known as one "half-life," 50% of the parent will have decayed to form an equal number of atoms of daughter. If the half-life of radioactive parent X is known, the age of a sample containing X its daughter isotope Y can be determined from the ratio of parent to daughter.

be calculated, because (as shown in Figure 2.7) it is equal to the sum of the remaining parent atoms (N) plus the new daughter atoms (D) radioactively formed, that is,

$$N_0 = N + D.$$

And because some of the daughter may also have been present at t = 0—for geologic samples, the time of crystallization of an igneous rock or of precipitation from solution of a sedimentary mineral—we may also have to make corrections for this. Fortunately, such corrections are not too complicated and as a result there are now known a number of radioactive isotopes that can be used effectively to date ancient rocks and minerals (Table 2.1).

Four main factors determine whether a radioactive element can be used in this way in the science of **geochronology:**

1. The decay constant (in essence, the half-life) must be known precisely and accurately.
2. The half-life must be geologically long (millions to billions of years) so that amounts of both the parent and daughter can be measured.
3. The isotopes must be of elements that occur commonly in enough abundance in rocks and minerals so they can be analyzed precisely and accurately.
4. The system analyzed (usually individual mineral crystals) must remain closed to loss or gain of parent or daughter from its surroundings.

It is crucial to also ask what actually determines t = 0, the time when the radioactive clock starts ticking? For isotopes such as carbon-14, produced by cosmic ray bombardment of nitrogen in the upper atmosphere, the clock starts ticking as atoms of ^{14}C are incorporated into living organisms. When an organism dies it stops adding ^{14}C and the clock begins to wind down—the longer the time since it died, the more half-lives pass and the less ^{14}C remains. Carbon-14 is useful for dating historical or archeological artifacts but since its half-life is only 5,730 years, too little of it remains to measure after about 50,000 years so it is useful only for dating the most recent events of the geologic past.

TABLE 2.1

Isotopic systems used commonly for dating in geology.

Parent	Daughter	Decay	Half life
^{40}K	^{40}Ar	electron capture	1300 Ma
^{238}U	^{206}Pb	series	4,470 Ma
^{235}U	^{207}Pb	series	704 Ma
^{232}Th	^{208}Pb	series	14,000 Ma
^{87}Rb	^{87}Sr	β	48,800 Ma
^{147}Sm	^{143}Nd	α	106,000 Ma

Half-lives are given in Ma (millions of years). The decay type labeled "series" represents several consecutive decays, commonly of different types, for which the daughter isotope listed is the ultimate product; as a result of the multiple decays, in these series there are large mass differences between parent and daughter.

In contrast with ^{14}C, the half-lives of isotopes used to date geological events are exceedingly long (Table 2.1)—hundreds or thousands of *millions* of years (Ma)—atoms made in stars before the Earth even formed. For these atoms the isotopic clock has been ticking a long time, so they are used to date events when parent and daughter isotopes come to be separated from each other or when minerals with different parent to daughter make-ups form in the same setting.

Consider, for example, the potassium-argon isotopic system. An isotope of potassium, ^{40}K, decays radioactively to an isotope of argon, ^{40}Ar (via electron capture—a process essentially opposite to β decay). Potassium is an element present commonly in minerals whereas argon is a nonreactive, inert gas. When potassium-bearing minerals crystallize from a molten mass of rock (a **magma**) they usually contain virtually no argon so the amount of radioactively produced ^{40}Ar that accumulates in the mineral relative to the amount of undecayed ^{40}K can be used to determine the mineral's age, the time since it crystallized.

A radioactive isotope of rubidium, ^{87}Rb, decays to a stable isotope of strontium, ^{87}Sr, by β decay (Table 2.1). Both rubidium and strontium are likely to be present in a magma, so any crystallizing mineral containing ^{87}Rb will also contain ^{87}Sr. Although some atoms of the daughter, ^{87}Sr, are usually present in a magma when it crystallizes, these can be accounted for by analyzing two or more minerals that have different parent to daughter ratios. If it is assumed that the isotopes of strontium are well mixed in a magma from which minerals crystallize, then the abundance of ^{87}Sr relative to one of the stable isotopes of strontium, such as ^{86}Sr, will be a constant. This, "the initial ratio," is shared by the magma and any minerals that crystallize from it at any given time. If several minerals crystallize at a given time with a particular initial ratio of ^{87}Sr/^{86}Sr, then at some later time the minerals with higher parent to daughter ratios will have higher ^{87}Sr/^{86}Sr ratios. All of the minerals that crystallize at that time will lie on a straight line in a plot of ^{87}Sr/^{86}Sr against ^{87}Rb/^{86}Sr, called an **isochron** (Figure 2.8), the slope of which is proportional to the age of their crystallization and the intercept of which on the ^{87}Sr/^{86}Sr axis gives the initial ratio.

The precise age of crystallization of igneous rocks can be dated quite easily because it represents an instant in geologic time. Other geologic events can also be dated by radioactivity. Metamorphic rocks can yield ages that show the time when temperatures were last high enough to mix isotopes thoroughly. For instance, at high temperatures potassium-bearing minerals are effectively purged of radioactive

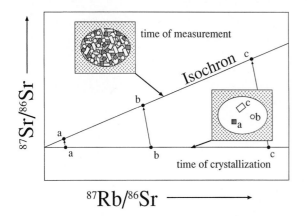

FIGURE 2.8 Isotopic dating using the isochron method accounts for daughter isotopes that may have been present at the time of the event dated. In the rubidium-strontium isochron method, the concentrations of isotopes are normalized to a stable isotope of strontium (^{86}Sr) because isotopic ratios can be measured more accurately than isotopic abundances. In the example shown here, the inset boxes illustrate a molten magma crystallizing into three different minerals ("a," "b," and "c") that have different parent to daughter concentration ratios. At the time of crystallization, the $^{87}Sr/^{86}Sr$ ratio of all three minerals is the same. Over time the $^{87}Sr/^{86}Sr$ ratio increases as a function of the parent to daughter ratio (which is proportional to $^{87}Rb/^{86}Sr$), and the line formed, called an isochron, has a slope proportional to the age of the rock formed from the three minerals.

potassium's daughter isotope, argon. But subsequent cooling through the temperature at which argon is retained in a crystal (the **closure temperature**) starts the isotopic clock. The specific temperature varies for different minerals and brings into question what event one actually is dating when the K-Ar system is used to analyze metamorphic rocks and minerals. In general, the dates record the time of metamorphism (but not necessarily its maximum amount which may have happened much earlier at a temperature far above the mineralic closure temperature). Nevertheless, careful use of isotopic data can be used to place limits on the thermal histories (the time-temperature relations) of metamorphic rocks and reconstruct the tectonic evolution of metamorphic terrains (Macdougal and Harrison, 1988).

Sedimentary rocks, on the other hand, are much more difficult to date by radioactive isotopes. This is because the process of sedimentation does not reset isotopic clocks—the parent-daughter decay systems in individual mineral grains are not affected when the grains weather out of the igneous or metamorphic rock in which they formed and are transported, deposited, and lithified in a sedimentary rock such as a sandstone or shale. There are, however, some exceptions. Certain minerals form in place while a sediment solidifies (that is, are **authigenic**), and these may be suitable for isotopic dating. A good example is glauconite, a potassium-bearing mineral related to micas which forms often in the subsurface layers of marine sands and can be used to date the lithified sediment.

Problems with Isotopic Dating

Although isotopic dating has been used quite successfully to establish the age of the Earth and to calibrate the planet's geologic history, it is not straightforward and simple because it usually involves a number of assumptions, some already discussed. Two of these deserve additional mention.

The first has to do with the decay constants (λ) of the widely used isotopic systems (Table 2.1) All have been determined with great accuracy and precision. Nevertheless, the validity of isotope dating hinges on the idea that "λ" is indeed constant, that it does not change in minerals and over time as factors such as pressure, temperature, and chemical environment change. A great number of experiments designed to test this assumption have confirmed the constancy of "λ," as does the fact that use of different isotopic systems to date a single rock sample give the same age.

The constancy of "λ" is, perhaps, not surprising, Radioactive decay is a nuclear process and although the electron shells that make up the outer part of an atom can be involved in chemical reactions and might be slightly compressible under high pressures, the nucleus is unaffected. Moreover, the energies involved in altering the nucleus are orders of magnitude greater than those of chemical processes. We therefore can have confidence that the decay constants of radioactive isotopes are independent of whether the isotopes occur in the form of atoms, molecules, or in a fluid or mineral.

A second concern is whether an isotopic system studied has remained closed and can therefore give a valid date. Processes such as weathering, metamorphism, and hydrothermal alteration can result in loss of parent or daughter isotopes from some systems. For example, because argon is an inert gas it is particularly susceptible to being lost from minerals when they are heated. Although heating to high temperatures may drive all argon out of the system and thus reset the isotopic clock (which can then be analyzed to give an age of metamorphism), more often only part of the argon is lost so that the $^{40}K/^{40}Ar$ ratio gives an age that cannot be clearly interpreted. Even elements as similar as rubidium and strontium have somewhat different chemistries (different solubilities, for example) which can lead to one or the other being removed or added to a mineral during weathering.

An effective way to circumvent problems such as these is to use isotopic systems that are composed of elements less mobile and less susceptible to being lost or gained. A good example is the system made up of the two rare earth elements samarium (Sm) and neodymium (Nd) which have very similar chemical behaviors. As shown in Table 2.1, radioactive ^{147}Sm converts to ^{143}Nd by α decay and has an exceedingly long half-life. The Sm-Nd method is therefore particularly useful for dating very ancient rocks, even those somewhat metamorphosed. However, the geochemical similarities between samarium and neodymium also limit the usefulness of the system since they prevent development of large variations of parent to daughter ratios and this, coupled with the long half-life of ^{147}Sm results in the Sm-Nd system having a very restricted range of measureable isotopic ratios.

DATING THE TIMELINE OF LIFE'S HISTORY

Soon after the discovery of radioactivity near the turn of the last century, it was recognized as a potential tool to date the geologic record (see, for example, Boltwood, 1907). By this time a rather detailed geologic time-scale based on biostratigraphic principles had been established and Darwin's work had shown that life evolved through time. The task now was to use the methods of radioactive dating to tie absolute ages to the already established stratigraphic record and by inference to the history of biologic evolution shown by fossils.

The most obvious problem was that (for the reasons discussed earlier) few sedimentary rocks could be dated directly. To work around this problem, ages were determined of rocks that could be dated directly—igneous rocks, for example, lava flows, **dikes** (sheets of rocks solidified from magma intruded into earlier formed rock units), and **plutons** (masses of rock solidified at depth)—and their spatial relations to sedimentary rocks used to bracket the age of fossil-bearing sediments. For this purpose lava flows are particularly useful because they spread and solidify across a preexisting surface (and are therefore termed **conformable** with the underlying surface). By the Law of Superposition, if a lava flow is deposited on top of a sedimentary rock unit then it is younger than that rock unit and any fossils it contains. The reverse, of course, also holds: Sedimentary units deposited on the top surface of a lava flow—or at any other level in a succession of sediments deposited above that flow—must be younger than the underlying lava. Cross-cutting bodies of rocks such as intrusive dikes and plutonic masses are also useful because any such body must be younger than the preexising rocks it intrudes (Figure 2.9) and of course must be older than any sedimentary rocks deposited (unconformably) above it in a sedimentary succession.

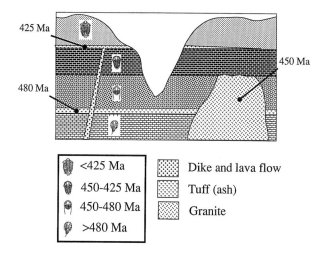

FIGURE 2.9   Isotopic dating of cross-cutting igneous rocks can be used to bracket the ages of associated sedimentary strata and the fossils they contain. In the example shown here in a vertical section through a succession of strata, the lowest unit of the sequence (#1, denoted by white bricklike symbols) was deposited first, followed by a layer of ash (unit #2) dated to be 480 million years (Ma) old, then strata of the immediately overlying (coarsely stippled) unit (#3). This three-unit sequence was intruded by a massive plutonic body of granite, dated to be 450 million years old, and the exposed surface (the top of unit #3 and the upper part of the pluton) was somewhat eroded. On this erosional surface a fourth unit (#4, black bricklike symbols) was deposited which was then overlain by a lava flow and intruded by an associated igneous dike (unit #5), dated to be 425 million years old, above which was later deposited the final unit of the sequence (#6, finely stippled pattern). The uppermost surface of the six-unit sequence was then deeply eroded to produce rolling topography and an incised valley. Unit #1 underlies unit #2, so it and its contained fossils are older than 480 million years, the age of unit #2. Unit #3 overlies unit #2 and so is younger than 480 million years, and is intruded by the granitic pluton and so is older than 450 million years; fossils in unit #3 are therefore between 480 and 450 million years old. Unit #4 overlies the granitic pluton so is younger than 450 million years, and is overlain by unit #5 and cross-cut by its associated dike so is older than 425 million years; fossils in unit #4 are therefore between 450 and 425 million years old. Unit #6 overlies unit #5 so the fossils it contains are younger than 425 million years. By this method, age ranges of index fossils (commonly a few million years) can be determined and the fossils themselves can then be used for dating strata even in areas distant from directly dated igneous rocks.

Using these principles, the modern geologic time-scale was developed. As we will see later, it continues to be modified and refined.

In addition to lavas, dikes, and plutons, one other type of igneous rock unit is particularly useful in radiometric dating, lithified layers of volcanic ash known as **tuffs**. Because ash is of igneous origin it can be dated readily, and because large eruptions spread ash over very wide areas, a tuff horizon effectively denoting a mere instant in geologic time can serve as a widespread **chronostratigraphic** marker bed, extremely useful for sorting out the time-relations among geologic units with which it is associated.

Paleomagnetism: An Additional Dating Tool

The customary methods of geochronology have now come to be well established: Lavas, dikes, plutons, tuffs, and similar igneous rocks are dated directly by isotopic methods; ages are then assigned to sedimentary strata based on their spatial relations to the dated igneous rocks; and by use of index fossils and the principles of biostratigraphy the sedimentary strata are used to assign ages to fossil-bearing rocks elsewhere, even if they are not closely associated to directly dateable igneous units. In recent decades yet another important dating tool, paleomagnetism, has been developed.

Anyone who has used a compass while hiking is well aware that the Earth has a strong magnetic field. The field is generated by the movement of fluids in the planet's molten iron-nickel outer core and can be thought of as a shell of lines around the globe that link the Earth's magnetic poles from south to north in the direction a compass needle points. Rocks that contain magnetic minerals commonly preserve a record of the magnetic field that existed when they formed. When igneous rocks pass through what is known as the **cooling temperature** as they crystallize from molten magma, magnetic minerals come to be aligned parallel to the magnetic field. And as magnetic minerals grains are deposited in sediments they tend to orient themselves parallel to the magnetic field in much the same way as a compass needle.

While today all compass needles point north rather than south, the opposite has been true repeatedly in the geologic past. The motions in the fluid core are such that the direction (polarity) of the magnetic field flips back and forth from time to time, the north magnetic pole switched to the south geographic pole then later back to the normal configuration. Both configurations are stable in the sense that once in place they last for thousands or millions of years, and the reversals happen abruptly and are episodic rather than repeating in a regular, rhythmic fashion.

Measurement of the magnetic field directions preserved in dated rock sequences has led to construction of a global paleomagnetic time-scale, the most recent 55 million years of which is shown schematically in Figure 2.10. The field direction can be used to label strata in much the same way as can index fossils, but in this case the labeling is binary—the field is either "normal" or "reversed." Used together with isotopically dated horizons and biostratigraphy, the paleomagnetic time-scale improves the resolution of geologic dating by providing ages for rock units present in intervals not dateable by isotopic methods or barren of index fossils.

The Major Divisions of Geologic Time

By international agreement, the Earth's geologic past is divided into two great eons—the **Precambrian Eon**, extending from the formation of the planet

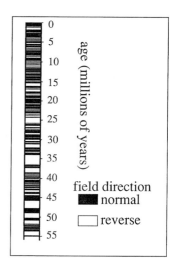

FIGURE 2.10 The paleomagnetic time scale for the most recent 55 million years of Earth history. The direction of magnetism can be determined in many different types of rocks and together with dated reference points used to provide detailed dating of geological sequences.

4.5 billion years ago to the first appearance in the rock record of fossils of hard-shelled animals (clams, snails, trilobites, and so forth) about 543 million years ago; and the **Phanerozoic Eon**, spanning the time from the end of the Precambrian to the present. Further subdivision of geologic time is based on events or markers in the rock record, for the Phanerozoic most commonly the appearance or extinction of specific groups of organisms. The isotopically determined absolute ages of these events shows life's evolution is neither rhythmic nor periodic and allows us to evaluate the *rates* at which evolution occurs (a topic discussed in Chapter 3).

Although the Earth was formed about 4.5 billion years ago, the oldest rocks dated so far, metamorphosed sediments from northwestern Canada, are only 4 billion years in age (Bowring *et al.*, 1989). No rocks as old as the Earth itself have been found and probably none still exist because the early planet was very active, much of it molten, and any solid rocks formed would have long since been eroded away. But rocks dating from the time of Earth's genesis have been found on the Moon where they have been preserved because of the Moon's more placid geologic history.

Life was already thriving on Earth by 3.5 billion years ago, the age of the oldest known fossils (tiny bacteriumlike microorganisms; Schopf, 1993) and may have existed as early as 3.8 billion years ago (Mojzsis *et al.*, 1997) or even earlier. Fossils of large many-celled organisms are earliest known from rocks about 600 million years old, formed near the close of the Precambrian after nearly 90% of Earth's history had already passed. Both the rock record and the fossil record of the Phanerozoic (from Greek, meaning "visible life"), the most recent 543 million years of Earth history, are much better known than those of the Precambrian. The resolution of the time scale improves markedly as it approaches the present—events can be dated with increasing accurately which permits the stratigraphy to be subdivided much more finely.

DATING THE FOSSIL RECORD: TWO CASE STUDIES

To illustrate how various dating techniques can be integrated and, especially, to highlight the usefulness of absolute ages provided by radioactive isotopes, we turn now to two case studies, one from near the beginning of the Phanerozoic Eon, the other from the recent geologic past.

The Precambrian/Cambrian Boundary

The oldest **geologic period** of the Phanerozoic is known as the **Cambrian** and extends from about 543 to 505 million years ago, its beginning marked in the rock record by the abrupt, "explosive" appearance of abundant shelled multicellular animals. Earlier, during the Precambrian Eon, life's history was dominated by much simpler forms of life—protozoa, single-celled algae, and diverse types of primitive bacteriumlike microorganisms.

But the shelled fossils of the basal Cambrian are not the oldest animals known—in Australia, Russia, Namibia, and several other regions, strata that immediately underlie Cambrian sediments contain easily recognizable flattened impressions of a latest Precambrian (**Vendian**-age) assemblage of fairly large soft-bodied animals known as the **Ediacaran Fauna**. The exact biologic relations of the dozen or so types of organisms in the fauna have been subject to considerable debate. Although some forms have been compared with jellyfish and segmented worms, others seem unlike any organism living today and most bear little resemblance to fossils abundant in the overlying Cambrian.

Until the past few years it seemed likely that the Ediacaran organisms are separated from the younger and more familiar fossils of the basal Cambrian by a substantial gap in time of probably many tens of millions of years (Figure 2.11. left), but there were no solid isotopic dates from any of the older Ediacaran-age successions so there was no way to know. This uncertainty raised important evolutionary questions. If the Ediacaran organisms were direct ancestors of the Cambrian fauna and separated from it by a large time gap, then why was there no fossil record in intermediate-age rocks to connect the two? Some argued that the missing record must exist but simply hadn't yet been found, whereas others pointed to differences in the make up of the two faunas and argued that the Ediacaran assemblage was an evolutionary dead-end, a failed "experiment" from an earlier epoch of evolutionary history.

Within the past few years, however, new evidence has shown that the gap between Ediacaran and Cambrian times was actually very short, and as the gap all but vanished evolutionary uncertainty diminished. Recently reported uranium-lead isotopic dates on grains of the mineral zircon extracted from tuff beds in Namibia show there is almost no time gap between the Ediacaran and Cambrian faunas (Grotzinger *et al.*, 1995). Indeed, precise dating shows that the youngest Ediacaran fossils predate the beginning of the Cambrian by a scant 6 million years (Figure

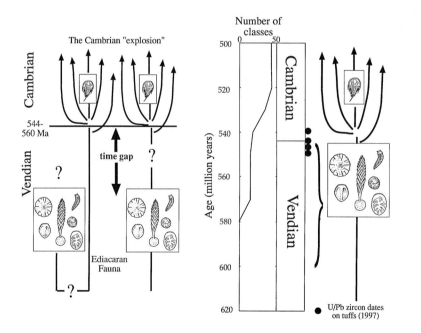

FIGURE 2.11 The use of isotopic dating to determine the evolutionary significance of the Vendian-age Ediacaran Fauna (based on Grotzinger *et al.*, 1995). Prior to the past few years (left), there was widespread debate about the evolutionary relations between the Ediacaran Fauna and lowermost Cambrian fossils, in part because of a presumed time gap between the two biologic assemblages. New radiometric ages (right), determined using U-Pb (Table 2.1) isotopes in the mineral zircon extracted from tuff layers, shows that rocks containing Ediacaran fossils are nearly as young as the oldest rocks of the Cambrian and increases the likelihood of an ancestor-descendant relationship between Ediacaran and Cambrian fossils.

2.11, right). This result strengthens the case for a direct ancestor-descendant relationship between Ediacaran fossils and the Cambrian fauna and shows exactly where in the geologic column intervening fossils should be sought.

Early Humans in East Africa

Among the most fascinating questions in science are those that center on the timing and nature of the origin of humans. Probably the richest finds of ancient hominids have been unearthed in East Africa where anthropologists and geologists have for many years carried out detailed studies of the rocks and the fossils and artifacts they contain.

The **East African Rift** is a long fertile valley created by plate tectonic movements that have pulled apart that region of the African continent. The same geologic processes have led to formation of volcanoes along the zone of rifting which over the past several million years have erupted frequently and spread ash over wide areas. The ash rapidly collects in basins and lakes along the Rift Valley where it hardens to distinct layers of tuff that are in part responsible for the fertility of the soils. For at least the past 5 million years environmental conditions in East Africa have been ideal for the evolution of hominids, a record of which is preserved in the Rift Valley sediments. Rarely, evidence of hominids is found even in the tuff layers, such as the famous footprints preserved in ash found by Mary Leakey in 1978 at Lateoli, Tanzania. Isotopically dated as about 3.5 million years old this ash layer provides the earliest evidence of hominid bipedality.

Isotopic dating, especially of tuff layers, has proven invaluable for sorting out early human history. Indeed, dates from tuffs interspersed between hominid-bearing horizons at sites such as Koobi Fora in northern Kenya (a stratigraphic column for which is illustrated in Figure 2.12) have enabled science to set minimum ages for appearance of such members of the human family as *Australopithecus africanus*, *Homo habilis*, and *Homo erectus* (McDougall *et al.*, 1980; McDougall, 1985) and to calibrate the rates of evolution of these and others of our ancient ancestors.

CONCLUSIONS

All natural sciences, but perhaps especially the geosciences including paleontology, rely heavily on an historical perspective. But although the histories of the Earth and life are preserved in the rock record, they are difficult to decipher because of the dynamic nature of the planet. Movements of massive crustal tectonic plates generate volcanoes, earthquakes, and uplift mountain ranges while weathering by wind, rain, and ocean waves continually wear down the planet's surface. As a consequence, the more recent record of rocks and the fossils it contains is better preserved, easier to decipher, and much more abundant than the ancient.

Confronted by difficulties such as these it is no wonder that pioneering workers such as Joly, de Buffon, and Kelvin were unable to estimate accurately the age of the planet. But their work was pivotal in establishing the enormity of geologic time which when finally recognized by the early 1900s spurred efforts to devise a global geologic time-scale. Bequerel's 1896 discovery of radioactivity provided the means, used over the present century to develop an array of increasing refined methods for isotopic determination of the absolute ages of minerals and the rocks they form. Coupled with use of index fossils and the principles of biostratigraphy, these advances have paved the way toward tracing the history of the Earth and of life into the distant geologic past.

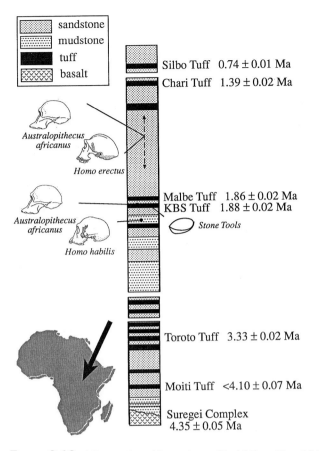

Figure 2.12   The stratigraphic section at Koobi Fora, East Africa, shows the temporal distribution of tuff layers and measured ages relative to fossil hominids and hominid artifacts (after McDougall *et al.*, 1980). These and data like them provide a strong scientific basis for deciphering the times of earliest appearance of human ancestors such as *Homo habilis* and *Homo erectus* and for understanding their rates of evolution.

Acknowledgments

I thank Charles Marshall, author of the following chapter of this volume, for his helpful discussions of some of the ideas and information presented here, and Bill Schopf for inviting me to contribute to this volume.

References

Boltwood, B. B. 1907. On the ultimate disintegration products of radio-active elements. Part II. The disintegration of Uranium. *American Journal of Science* **23**: 77-88.

Bowring, S. A., Williams, I. S. and Compston, W. 1989. 3.96 Ga gneisses from the Slave Province, N.W.T., Canada. *Geology* **17**: 971-975.

Dalrymple, G. B. 1991. *The Age of the Earth* (Stanford, CA: Stanford University Press), 474pp.

Davidson, J. P., Reed, W.E. and Davis, P.M. 1997. *Exploring Earth* (Upper Saddle River, NJ: Prentice Hall), 477 pp.

Gould, S. J. 1987. *Time's Arrow, Time's Cycle: Myth and Metaphor in the Discovery of Geological Time* (Cambridge, MA: Harvard University Press), 222 pp.

Grotzinger, J. P., Bowring, S. A., Saylor, B.Z. and Kaufman, A. J. 1995. Biostratigraphic and geochronologic constraints on early animal evolution. *Nature* **270**: 598-604.

Hawking, S. W. 1988. *A Brief History of Time, From the Big Bang to Black Holes* (London: Bantam Press), 198 pp.

Hutton, J. 1788. Theory of the Earth. *Transactions of the Royal Society of Edinburgh* **1**: 209-305.

Lyell, C. 1877. *Principles of Geology,* 11th edition (New York: D. Appleton), Vol. 1, 671 pp., Vol. 2, 652 pp.

McDougall, I. 1985. K-Ar and ^{40}Ar/^{39}Ar dating of the hominid-bearing Pliocene-Pleistocene sequence at Koobi Fora, Lake Turkana, Northern Kenya. *Geological Society of America Bulletin* **96**: 159-175.

McDougall, I. and Harrison, T. M. 1988. *Geochronology and Thermochronology by the ^{40}Ar/^{39}Ar method. Oxford Monographs on Geology and Geophysics, No. 9* (Oxford, UK: Oxford University Press), 212 pp.

McDougall, I., Maier, R., Sutherland-Hawkes, P. and Gleadow, A.J.W. 1980. K-Ar age estimate for the KBS tuff, East Turkana, Kenya. *Nature* **284**: 230-234.

Mojzsis, S. J., Arrhenius, G., McKeegan, K. D., Harrison, T. M., Nutman, A. P. and Friend, C. R. L. 1996. Evidence for life on Earth before 3,800 million years ago. *Nature* **384**: 55-59.

Schopf, J. W. 1993. Microfossils of the Early Archean Apex chert: New evidence of the antiquity of life. *Science* **260**: 640-646.

Further Reading

Ager, D. V. 1993. *The Nature of the Stratigraphical Record* (Chichester, UK: Wiley and Sons), 151 pp.

Dott, R. H. and Prothero, D.R. 1994. *Evolution of the Earth* (New York: McGraw Hill), 569 pp.

Eicher, D. L. 1976. *Geological Time* (Englewood Cliffs, NJ: Prentice Hall), 150 pp.

Holmes, A. 1927. *The Age of the Earth: An Introduction to Geological Ideas* (London: Harper), 80 pp.

Kirkaldy, J. F. 1971. *Geological Time* (Edinburgh, UK: Oliver and Boyd), 133 pp.

Stanley, S. M. 1989. *Earth and Life Through Time* (New York: W.H. Freeman), 687 pp.

CHAPTER 3

MISSING LINKS IN THE HISTORY OF LIFE

CHARLES R. MARSHALL[1]

Introduction

Ever since Darwin proposed his theory of evolution (or more correctly, theories; see Mayr, 1991) it has been assumed that intermediates now extinct once existed between living species. For some, the hunt for these so-called **missing links** in the fossil record came to be almost an obsession, a search for the key set of evidence thought needed to establish the veracity of evolutionary theory. Few modern paleontologists, however, search explicitly for ancestors in the fossil record because we now know that fossils can be used to chart the order of evolution regardless of whether they are directly ancestral either to extinct organisms or those living today.

This chapter provides an introduction to the way modern paleontologists use fossils to tell the course of evolution and in doing so addresses a number of interrelated questions: Can ancestors be identified in the fossil record? How common do we expect them to be? How do modern paleontologists trace the course of evolution without needing to know the ancestral forms? How fast can evolution happen—and how fast has it actually happened? Does evolutionary change take place fairly continuously or does it happen in bursts followed by quiescence? What do modern advances in **developmental genetics** tell us about the nature of the evolutionary transformations between species?

In light of the answers to these questions, the chapter concludes with a brief analysis of the contribution the fossil record has made to understanding the evolution of our own species, *Homo sapiens*, with special focus on the rate of evolution of the large human brain and the interesting question of the origin of language.

[1]Department of Earth and Space Sciences, Institute of Geophysics and Planetary Physics, and Molecular Biology Institute, University of California, Los Angeles, CA 90095.

EVOLUTION, ANCESTORS, AND MISSING LINKS

In Medieval times it was customary to organize the major groups of organisms into the "Great Chain of Being," a ladderlike array starting with forms of life that seemed simplest (at least to the naked eye) and building to the culmination of God's creation, *Homo sapiens*. But the Great Chain eventually fell apart when Charles Darwin and Alfred Russel Wallace codiscovered the theory of evolution which shows species are related by a branching pattern, a **phylogenetic** tree.

The theory of evolution led some to suggest that humans were derived from chimpanzees, the living primates seemingly closest to us (Figure 3.1A), a notion others found repugnant. (And one can imagine that the opposite pattern—a phylogeny like that shown in Figure 3.1B in which chimps evolved from humans— might be just as repugnant to chimpanzees!) Both these views are mistaken. Except under exceedingly rare circumstances, no living species is another's direct ancestor. Humans did *not* evolve from chimpanzees nor did they evolve from us. We both acquired our special characteristics after we diverged from our last common ancestor, a species that though certainly a primate was just as certainly neither chimpanzee nor human (Figure 3.1C).

The "Missing Link"

Though a living species is seldom the direct ancestor of another living species, all organisms are related by evolution so intermediates must exist. And if these links are missing among the living, the place to look is the fossil record. This logical byproduct of evolutionary theory was recognized as early as 1868 (only a few years after the 1859 publication of Darwin's *The Origin of Species*) by the great German zoologist and philosopher Ernst Haeckel who hypothesized the existence of a species of fossil human to bridge the evolutionary gap between humans and apes (though he mistakenly thought humans more closely related to the Asian orangutan

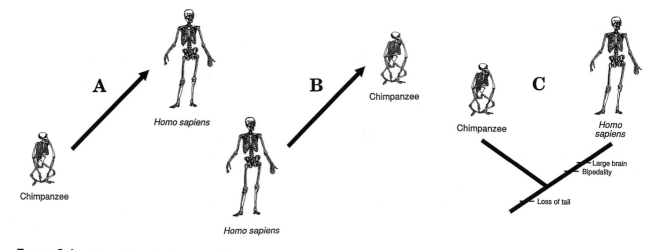

FIGURE 3.1   The relationship between chimpanzees and humans. Humans did not evolve from chimpanzees (A), nor did chimpanzees evolve from humans (B). Rather, chimpanzees and humans share a common ancestor and have acquired their own unique characteristics since divergence from that ancestor (C). Two of the unique characters on the human branch are an upright stance and a very large brain. These evolved after the loss of a tail which happened before chimpanzees and humans diverged from their last common ancestor.

and gibbon than chimpanzees, as we know today). Haeckel even proposed a double-barreled Latin name for this hypothetical missing link, *Pithecanthropus alalus*. (The scientific name of every species consists of two parts, a genus name, shared by a number of similar forms, and a species name, unique to that species within the genus. For his hypothetical missing link, Haeckel used the genus name *Pithecanthropus*—Latinized form of the Greek *pithekos*, ape, plus *anthropos*, human being, literally "ape-man"—and because he thought speech to be a key feature of humans that a missing link would have lacked he chose the species name *alalus*—the Latinized version of the Greek word *alalos*, "without speech.")

Haeckel's proposal captured the imagination of the Dutchman Eugène Dubois who became obsessed with finding the missing link. In 1891-92, Dubois found a few bones of a likely candidate at Trinil in Java (Figure 3.2). The fossil remains were too incomplete to allow him to determine whether the Trinil hominid was capable of speech, but a well preserved thigh bone showed that it walked upright, that is, had a **bipedal** stance. So though Dubois thought his find to be Haeckel's missing link, he nevertheless named it *Pithecanthropus erectus*, rather than *Pithecanthropus alalus* (and it has since been renamed *Homo erectus*).

The Trinil hominid is important because it helps us determine the sequence of events that led to development of modern humans. For example, even though *Homo erectus* was bipedal, the Trinil and other specimens of the species have smaller brains than modern humans. Evidently, the ability to walk upright evolved before the brain reached its modern size (Figure 3.3). The Trinil fossils are between 750,000 and 500,000 years old (see Chapter 2 for a discussion of the way fossils are dated), so Dubois' discovery was regarded as extraordinary since his fossil showed that hominids having an upright stance were already present at least half a million years ago (a trait that has since been traced to some two million years earlier, as discussed later in this chapter). This example illustrates one way fossils can be used to determine the *order* in which evolutionary inventions happen, but I now

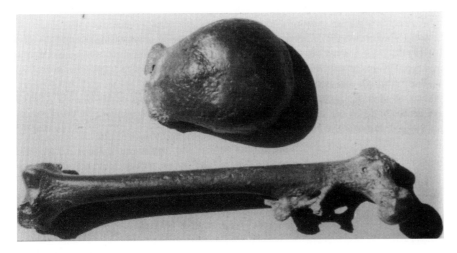

FIGURE 3.2 The "missing link," the fossil brain case and femur found in the late 1800s by Eugène Dubois at Trinil in Java, Indonesia. These bones were the first evidence of Ernst Haeckel's hypothesized missing link, *Pithecanthropus alalus* ("ape-man without speech"). But since Dubois had no way to determine whether this species could speak, he named his find *Pithecanthropus erectus* ("ape-man that walked erect") because the shape of the femur (the long leg bone, below) showed the species had an upright stance. These fossils of so-called Java man are now placed in the hominid species *Homo erectus*. (From Tobias, 1992, republished with permission.)

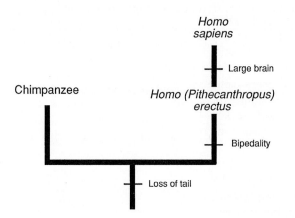

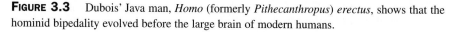

FIGURE 3.3   Dubois' Java man, *Homo* (formerly *Pithecanthropus*) *erectus*, shows that the hominid bipedality evolved before the large brain of modern humans.

want to show that fossils can also help us know the *course* of evolution, even if the fossils are not directly ancestral to any species living today.

Identifying Ancestors in the Fossil Record Is Difficult

Since Dubois' pioneering efforts, many new species have been discovered that appear more closely related to modern humans than the **great apes** (chimpanzees, gorillas, and orangutans). And though *Homo erectus* is a good candidate for an ancestor of modern humans, many of the others are not. A prime example is *Australopithecus boisei*, known from several East African sites about 2.5 to 1.0 million years in age, which lacks a number of important traits shared by modern humans and *Homo erectus*—such as a relatively large brain—and appears more distantly related to us than *Homo erectus*. But anatomical evidence shows that *Australopithecus boisei* is more closely related to humans than to any of the living apes, and for this reason it is classified officially as a **hominid**, unlike chimps or gorillas.

Other evidence shows that *Australopithecus boisei* was probably not ancestral to humans. For example, one trait of the fossil species is a prominent bony ridge on the top of the skull (Figure 3.4A), a **sagittal crest** where extraordinarily powerful jaw muscles attached. If we were to assume that *Australopithecus boisei* belongs in the lineage of our direct ancestors we would be forced to theorize that the lineage first evolved a sagittal crest which then was lost before *Homo erectus* arose (Figure 3.4B). The gain then rapid loss of such an evidently important feature is unlikely to have happened, so it seems far more plausible that *Australopithecus boisei* represents a separate branch of the evolutionary tree, a hominid not directly ancestral to humans(Figure 3.4C).

This brief analysis of where *Australopithecus boisei* belongs in hominid evolution illustrates one of the central tenets of modern paleontology: To be regarded as ancestral, a fossil is expected to lack unique characteristics (termed **autapomorphies**) that set it apart from other members of the lineage. Otherwise, one has to hypothesize that such traits were gained by the ancestor only to be immediately lost before the origin of the next descendent species. In some sense, a defining characteristic of ancestors is that they lack defining characteristics!

That a fossil species gains the status of "ancestor" chiefly on the basis of negative evidence, a lack of unique traits, cannot be counted on to always give the correct answer since there is always the possibility that a new-found fossil will have characteristics which would require the supposed ancestor to be assigned to a separate lineage. Moreover, many biological characteristics are not preserved in fossils (biochemical and behavioral traits, for example), so it is easy to image that were it possible to examine living representatives, a fossil species now regarded as an ancestor might actually be shown to belong to a different lineage. Given the ever-present possibility that new fossils with telling characteristics may be discovered and that many traits that define living species are not preserved in fossils, hypotheses of direct ancestor-descendant relations are always provisional.

Ancestors Are Not Important to Paleontologists

At first glance, the inability to establish firm ancestor-descendant relations might seem to cripple paleontology. After all, if the science cannot identify ancestors and their descendants with certainty, how can fossils contribute to understanding the course of evolution? The answer is simple: A fossil species does not need to be ancestral to any known species for it to contribute to understanding evolution's path. As an example, it would not matter to understanding of human evolution if a new-found exquisitely preserved specimen showed *Homo erectus* to belong to a evolutionary lineage separate from modern humans because the combination of

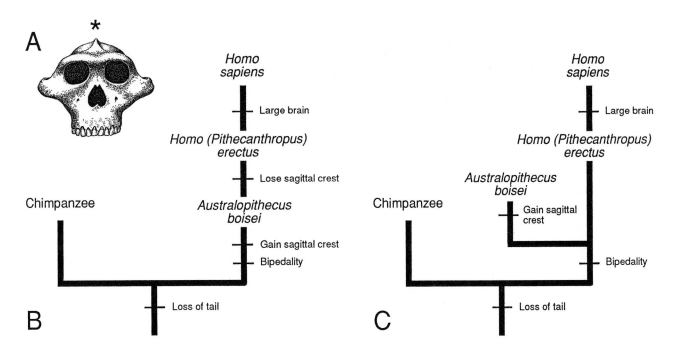

FIGURE 3.4   Not all fossil species are ancestors of living species. (A) A frontal view of a skull of *Australopithecus boisei* showing the pronounced sagittal crest (*) to which thick jaw muscles attached (modified after Walker and Shipman, 1996). (B) If we assume that *Australopithecus boisei* was directly ancestral to *Homo erectus* and *Homo sapiens*, then we must hypothesize that the sagittal crest was lost almost as soon as it was gained. (C) A simpler hypothesis is that *Australopithecus boisei* is related to the two species of *Homo* but belongs to a separate branch, now extinct, in which the sagittal crest was gained then never lost. This example illustrates the general rule that species having prominent unique characters (such as the sagittal crest) are most commonly not direct ancestors of species that lack such characters.

traits it possesses would still suggest that human upright stance evolved before large brain size. It is the specific combination of characteristics shown by a species, whether living or extinct, that allows us to infer the course of evolution. Paleontologists do not need to identify specific ancestors to trace evolution's path—all they need to identify are evolutionary innovations shared between species (**synapomorphies**). "Missing links" can fill in gaps without being actual ancestors.

Synapomorphies hold the key to unraveling life's evolutionary history, a principle illustrated by examining how the oldest known fossil bird, *Archaeopteryx*, has increased our understanding of the order in which features unique to modern birds appeared. *Archaeopteryx* is known from rocks formed late in the **Jurassic Period** of Earth history, some 145 million years ago, in Bavaria, Germany. Famous because of their beautifully preserved feathers, skeletons of *Archaeopteryx* closely resemble those of some dinosaurs, especially meat-eating **theropods** such as *Deinonychus* and *Velociraptor* (though some paleontologists think *Archaeopteryx* is more closely related to fossil relatives of crocodiles). In particular, like dinosaurs *Archaeopteryx* has a small brain case, three unfused fingers each with its own claw, teeth, a relatively small breastbone (**sternum**), and a long tail. Regardless of whether *Archaeopteryx* is in the direct line of descent to modern birds or represents a separate branch of the evolutionary tree of birds, its **morphological** attributes indicate that in bird evolution feathers evolved before the fusion of fingers, loss of claws and teeth, enlargement of the sternum, and shortening of the tail (Figure 3.5). Regardless of its specific ancestor-descendant relations, *Archaeopteryx* is exceptionally informative about the course of evolution.

But while it sometimes seems easy to trace evolution's path, this does not mean that the history of a lineage was preordained to continue down the path it did. Diagrams like Figure 3.5 should not be misread to imply an inevitability to the course of evolution.

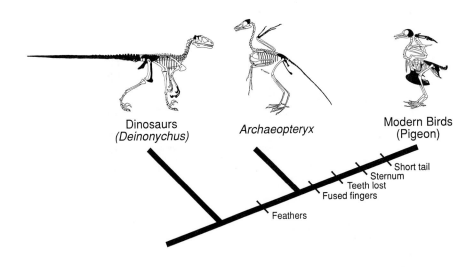

FIGURE 3.5   The spectacular birdlike fossil *Archaeopteryx* helps us determine the order of evolution of innovations unique to birds. Whether *Archaeopteryx* was directly ancestral to birds or not, the combination of characters it possesses show that feathers evolved early in avian history and that other traits of modern birds evolved later. (Drawings modified after Colbert, 1980.)

But Ancestors Should Exist in the Fossil Record: How Common Are They?

Though it may be difficult to be sure that a fossil "ancestor" is actually an ancestor, evolution tells us that all living (and extinct) species did indeed evolve from earlier ancestral species. Moreover, we can expect the ancestors to be preserved in the fossil record at least occasionally. What proportion of fossil species now known might actually be direct ancestors either of other fossils or of species living today? A computer simulation of the fossil record can be used to answer the question. R. D. Martin (1993) estimates that only about 3% of all primate species that ever existed are represented in the fossil record as now known. On the basis of this estimate of the incompleteness of the known fossil record, Tavaré and Martin performed a simple simulation to mimic the effects that this incomplete paleontological sampling has on knowledge of the primate evolutionary tree (Martin 1993). They constructed a hypothetical evolutionary tree (Figure 3.6, above), a computer-generated **phylogeny** made up of 333 "primate species," then sampled at random 3% of the extinct forms (Figure 3.6, below). At this level of sampling, not a single direct ancestor-descendant pair was detected and only two of the 10 randomly selected

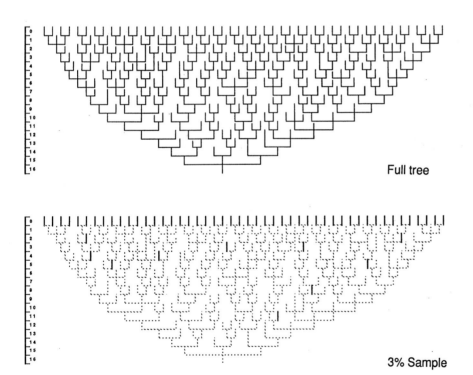

Full tree

3% Sample

FIGURE 3.6 Computer simulations of the fossil record show that when relatively few species are preserved, ancestral species are rarely present. (Above) A simulated phylogenetic tree in which primate species are represented by vertical lines and their evolutionary relations by horizontal lines. "Living" species are at the top of the tree (time slice 0) whereas species terminating in time slices 16 to 1 are "extinct." (Below) If only a small percentage of past species are preserved and thus available for paleontologic sampling, few ancestors can be detected. For example, if only 3% of extinct species of the full tree are sampled (the 10 solid vertical bars), no direct ancestor-descendant pairs are revealed and only two of the ten fossil species found are even indirectly ancestral to other fossil species. (Modified after Martin, 1993.)

fossil species were even indirectly ancestral to any of the others selected! So, with an incomplete fossil record, direct ancestors can be expected to be rare and even indirect ancestors uncommon.

Mathematical methods can also be used to estimate the number of direct and indirect ancestors that should be present in the fossil record. For example, M. Foote (1996) derived some fairly simple equations to estimate the proportion of preserved fossil species that should be either directly or at least indirectly ancestral to other preserved species. Figure 3.7 is a graph of the results of one of his models. As might be expected, the graph shows that the probability of any two known fossil species being directly related as ancestor and descendant increases with the proportion of species preserved. The graph also confirms Tavaré and Martin's conclusion that direct ancestors (Figure 3.7; lower curve) should be much rarer than indirect ancestors (Figure 3.7; upper curve).

These studies show that ancestors *should* be present in the fossil record. But though they no doubt exist they are still difficult to identify because fossil organisms are never perfectly preserved. There is no good way to evaluate what fraction of original morphology is faithfully preserved in most fossils, but it certainly is far less than one might guess from museum exhibits where only the most complete fossil specimens are displayed. And the quality of preservation varies greatly depending on the organism—clam shells, for example, are often well preserved whereas skeletons of **vertebrates** (animals having backbones) are seldom preserved intact and animals that lack hard parts, such as worms and slugs, are rarely preserved at all.

An indication of the quality of preservation for at least one group of vertebrates can be gleaned from P. Dodson's (1990) summary of the completeness of dinosaur remains. Of the 285 genera of dinosaurs thus far discovered (composed of 336 species), almost half are known from but a single bone or skeletal fragment! Only 20%, one of every five genera, are known from complete skulls and skeletons.

It stands to reason that if some fossil genera are known from only a single bone, there must be others not (yet) known at all. Though there is no way to know

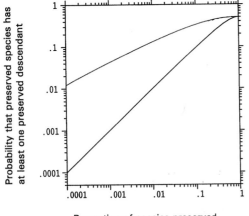

FIGURE 3.7 How common in the fossil record are species ancestral to other fossil species? This graph shows the probability of a fossil species being directly ancestral (lower line) or at least indirectly ancestral (upper line) to another fossil species. (Modified after Foote, 1996.)

for sure how many different types of dinosaur-sized vertebrates the Earth's ecosystem once supported, Dodson used the number of genera of large mammals today to estimate that over their 150-million-year-long history a total of between 900 and 1,200 dinosaur genera roamed the planet. The 285 genera now known to science represent about 25% of the total of which only one in five—a scant 5%—of all that probably ever lived are known from complete skulls and skeletons.

How many dinosaur genera might be ancestral to other genera of dinosaurs? Though technically, a genus cannot give rise to another genus (even though a *species* of a genus can give rise to the founding species of a different genus), if we let the number of dinosaur genera proxy for their species we can estimate the proportion of ancestors among the dinosaurs now known (a method permissible since on average the described genera consist of only 1.2 species and 86% of but a single species). Let's first consider the 60 or so dinosaur genera known from complete skeletons. According to Foote's graph (Figure 3.7), about 5% (3 genera) should be directly ancestral, and 30% (18 genera) at least indirectly ancestral to other genera known from complete remains. For all known genera, regardless of the quality of preservation, 20% should be direct ancestors and 45% at least indirect ancestors (though because so many genera are known from only partial remains, an even larger percentage might seem to be ancestral). Even in the scanty fossil record now known, ancestors among dinosaurs should be fairly common.

THE MODE OF EVOLUTIONARY CHANGE

Some 25 years ago it was widely (but not universally) believed that evolutionary change happened more or less continuously over time as a result of small changes accumulated in species over many generations. This mode of evolution is known as **phyletic gradualism** (Figure 3.8, left). Abrupt species-to-species changes seen in the fossil record were blamed on its incompleteness and explained away as gaps to be filled by fossils yet to be discovered. This view was challenged in 1972 by Niles Eldredge and Stephen Jay Gould who proposed the theory of **punctuated equilibrium** (Figure 3.8, right) according to which gaps in the fossil record are real— rather than artifacts of incomplete preservation—caused by brief bursts of evolutionary change, especially during the formation of new species (**speciation events**), that "punctuate" long periods of equilibrium when there is little or no change (**stasis**).

Which of these hypotheses is more nearly correct? The new idea, Eldredge and Gould's proposal of punctuated equilibrium, sent paleontologists scurrying to test it against the fossil record. Now after more than 60 studies over a quarter-century it appears that both modes of evolution happen as do practically all variations between (Erwin and Anstey, 1995). For example, as illustrated in Figure 3.9, strong evidence for phyletic gradualism is shown by P. D. Gingerich's 1974 analysis of the change in size of small **Eocene**-age mammals (extinct **condylarths**) over a period of about 5 million years (in which he circumvented the lack of complete skeletal remains by measuring the size of the first upper molar as a proxy for body size). But studies also show support for punctuated equilibrium. B. A. Malmgren *et al.* (1983) found this model to fit the rapid shift in size shown during the evolution of the foraminiferan *Globorotalia plesiotumida* to its larger descendant, *Globorotalia tumida* (Figure 3.10). **Foraminifera** are tiny single-celled marine organisms that because of their rich fossil record are ideal for this type of study. Interestingly, over time both of the species studied by Malmgren *et al.* fluctuate in size, so in their evolution we are seeing a mixture of punctuated equilibrium *and* phyletic gradualism.

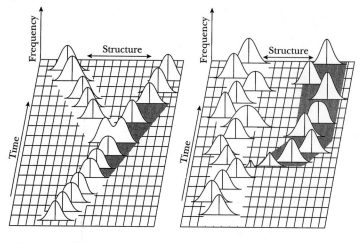

Phyletic Gradualism   **Punctuated Equilibrium**

FIGURE 3.8   Two models of modes of evolution, phyletic gradualism (left) and punctuated equilibrium (right). The bell-shaped curves show the range of variation of traits of each species at a particular time. In phyletic gradualism, no acceleration in the rate of change in size or shape is associated with the budding off of a new species, shown by the speciation (divergence) event near the center of the diagram. In contrast, in punctuated equilibrium rapid changes are associated with the origination of new species (shown beginning with a very small population, denoted by the small size of the bell-shaped curve at the lower middle of the diagram). (Modified after Conroy, 1997.)

Prior to the notion of punctuated equilibrium there was a tendency to interpret fossils in light of entrenched theory and attribute discrepancies between theory and observation to the incompleteness of the preserved record. So one of the most important aspects of the new idea is that it forced paleontologists to look critically at the fossil record and develop news tools to measure rates of change.

The lack of input from the fossil record to theory is well illustrated in Figure 3.8. The diagram that represents punctuated equilibrium does not show perfect stasis during periods of equilibrium but instead realistically shows species changing a little from one time to the next and therefore fairly faithfully reflects what is known from the fossil record (as also illustrated, for example, in Figure 3.10). In contrast, the diagram in Figure 3.8 that represents phyletic gradualism shows a completely unvarying rate of change, despite the fact that even the best examples of phyletic gradualism (for example, Figure 3.9) show fluctuations in these rates.

WHAT DRIVES EVOLUTION?

Before we can discuss how fast evolution proceeds we must learn a little about what makes evolution go. There are several important ingredients (Marshall, 1995), two in particular. First, evolutionary change can result from environmental conditions that favor individuals with specific characteristics over others that lack those characteristics—that is, **natural selection** can drive evolution (though chance, also, can play a role). If humans are responsible for the selective pressure, as when cows are chosen for breeding based on milk production, we call this **artificial selection**.

But by itself, natural selection is not enough to make evolution go. The second prime ingredient is **heritability**, because unless the "good" (selected)

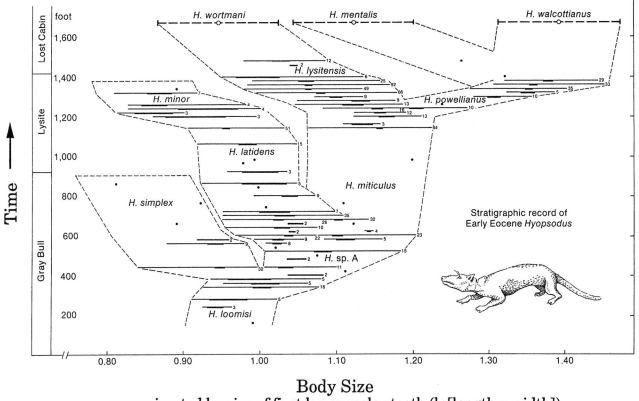

FIGURE 3.9 Evidence of phyletic gradualism in the evolution of molar tooth-size (a proxy for body size) in species of the Eocene condylarth mammal *Hyopsodus*. Fossils were recovered from 46 stratigraphic horizons (denoted by the horizontal lines and dots) of three rock formations (named at the left) that span about 5 million years. Each horizontal line shows the range of size for the number of specimens indicated. (Modified after Gingerich, 1974.)

characters are passed on to the next generation, natural selection would not lead to evolutionary change. To have an effect, the characteristics must be at least partly heritable.

Figure 3.11 shows the relation between selection and heritability. The top graph shows the distribution of milk yields for a **population** of cows in which most are about average, some produce much milk, some hardly any. From this population the farmer selects only cows with the highest milk yields to breed the next generation. The difference between the average milk yield of the parent population and the part of it chosen by the farmer as the breeding population is a measure of the strength of the (artificial) selection. The figure shows that in the next generation the average milk yield increases, but is less than that of the breeding population (the stippled region in the upper part of Figure 3.11). The reason for the incomplete response to selection is that milk yield is not completely heritable—some daughters give less milk than their mothers, some give more. But quite clearly, milk yield is at least partly heritable because if it were not, selection would have no effect. So, the **response to selection** is related both to the strength of the selection (the **selection differential**) and heritability. The higher the selection differential and the higher the heritability, the faster evolution will go.

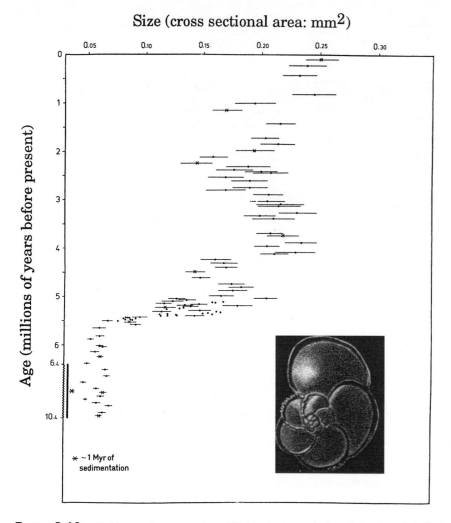

FIGURE 3.10   Evidence of punctuated equilibrium in the evolution of the size of shells in fossil protozoans (foraminiferans). Minor size fluctuations are punctuated by a rapid shift in size across the evolutionary transition from a smaller species (*Globorotalia plesiotumida*) to the larger descendant species (*Globorotalia tumida*). (Modified after Malmgren *et al.*, 1983, and Buzas *et al.*, 1987).

Is There Evidence for Selection?

Studies of evolution have provided reams of evidence for how natural selection acts (for example, Darwin, 1859; Endler, 1986; Weiner, 1994). Here I mention just one example, the strong selection against both light and heavy birth weights in humans (Figure 3.12)—babies especially underweight or overweight at birth have a very high rate of mortality. In this example there is selection against deviations from an average, a type of selection which operates *against* evolutionary change called **stabilizing selection**.

How Heritable Are Biological Characters?

The ability of natural selection to direct the course of evolution depends on the extent to which parents pass their characteristics to their offspring. And the degree of similarity between parents and their offspring depends on both genetic and

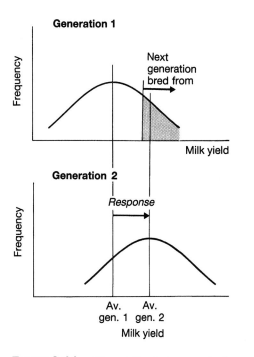

FIGURE 3.11 The relation between selection and heritability shown by breeding of milk cows. (Above) Of the total herd of cows in generation 1, only those having high milk yield (shown by the stippled region) are selected to breed to produce generation 2. (Below) That cows of generation 2 have a higher average milk yield than the total herd of generation 1 shows there has been a positive response to the artificial selection. But the magnitude of the response shows that milk yield is not completely heritable since the average yield of the daughter cows (generation 2) is lower than the average yield of their mothers (stippled region of generation 1). (Modified after Ridley, 1996.)

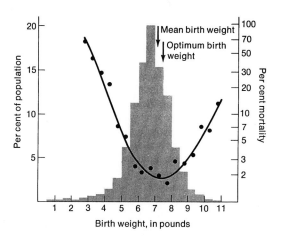

FIGURE 3.12 Evidence for natural selection in the birth weight of humans. Babies of both unusually low and high weights have a high rate of mortality whereas the lowest mortality (the optimum birth weight) is very close to the average (mean) weight. This is an example of stabilizing selection, a type of natural section that does not result in evolutionary change. (Modified after Futuyma, 1986, and based on data from Cavalli-Sforza and Bodmer, 1971.)

environmental factors (Futuyma, 1986; Ridley, 1996). The **genetic** factors fall in two classes depending on whether they are heritable—those that are passed to offspring make up the **additive** component whereas those not heritable comprise the **nonadditive** component. The heritability of a character, designated h^2 and defined mathematically as the proportion of the variation seen in a character that is due to additive genetic effects, has values that range from zero to unity. If a particular trait in a parent deviates from the average but in its offspring does not, the heritability of the trait is zero (and natural selection of that trait does not influence evolution). But if a character in a parent and its offspring deviate from the average by exactly the same amount, the heritability of the character is unity (and its selection has a maximal effect of evolution). It is common for traits in offspring to deviate from the average less than do the same traits in their parents so that traits' heritabilities, h^2, are greater than zero but less than one.

Careful studies have measured the heritabilities of many traits, especially those commercially important. Table 3.1 summarizes the results of a few of these studies. Some characters are almost completely heritable, such as the amount of white spotting in Friesian cows ($h^2 = 0.95$), whereas others have practically no heritability, such as the number of offspring per litter in mice ($h^2 = 0.15$) or the rate of conception in cattle ($h^2 = 0.01$). Many heritabilities fall in the mid-range of values, as each of us might expect from personal experience—brothers and sisters commonly show strong "family resemblance," an indication of substantial heritability, but we would be astonished if children grew to be exact copies of their parents.

RATES OF EVOLUTIONARY CHANGE

Now that we have explored some of the major factors that contribute to evolutionary change, let us consider the time scales over which change can be detected and how fast evolution proceeds.

How Fast Can and Does Evolution Go?

Three types of morphologic traits can be used to trace evolutionary change. One includes attributes that change in size, such as brain size or body size, characters known as **continuous traits**. A second includes traits that change in number—for example, the number of toes, ribs, or neck vertebrae—characters termed **meristic traits**. The third type are new characteristics, **neomorphic traits**, easy to spot because they are unlike the structures from which they evolve. Examples of neomorphs include the feathers of birds, modified by evolution from reptilian scales, or the sixth "finger" of a panda's paw which is actually an evolved wrist bone, anatomically not a finger at all (Gould, 1980).

The rates of change of continuous traits are the easiest to determine and are measured in units called **darwins**. One *darwin* (*d*) equals the rate of evolution that would produce a change in size by a factor of approximately 2.7 in one million years. (Technically, one *darwin* is a change in size per million years by a factor of e, which is equal to 2.718281828... and is the base of what are known in mathematics as **natural logarithms**.) *Darwins* permit us to quantify and compare rates of evolutionary change. For example, if in 10,000 years an animal the size of a mouse evolved to one the size of an elephant its rate of evolution would 400 d, whereas if the same size increase happened over 100,000 years the rate would be ten times slower, only 40 *d*.

Table 3.2 shows estimates of rates of evolution from artificial selection experiments, changes observed in species introduced to new geographic areas, and

TABLE **3.1**

Heritabilities of selected biological characters.

Trait	Heritability (h^2)
Cattle	
Degree of white spotting in Friesian cows	0.95
Milk yield	0.3
Conception rate	0.01
Sheep	
Length of wool	0.55
Body weight	0.35
Mice	
Tail length	0.6
Litter size	0.15
Chickens (white leghorn)	
Egg weight	0.6
Age at first laying	0.5
Egg production	0.2
Drosophila melanogaster (fruit fly)	
Abdominal bristle number	0.52
Humans	
IQ	0.34

Data from Futuyma (1986), except those for human IQ which are from Devlin *et al.* (1997).

studies of the fossil record. Rates of evolution can be extraordinarily high, up to 200,000 *d* in artificial selection experiments, but also extremely low, even zero for some fossil animals that show stasis, no measurable rate of change.

Perceived Rates Depend on the Time Scale of Observation

At first glance, Table 3.2 seems to show that the longer the time over which evolutionary change is measured, the slower its maximum rate. For example, the fastest rates measured in fossils over geological time are more than 3,000 times slower than the fastest rates measured for species observed as they colonize new areas. Though the measurements shown in the table are valid, the correlation is an artifact. Geological time and biological time are vastly different. Students of the fossil record almost always can only measure average rates of evolution over times that

TABLE 3.2

Summary of rates of evolution.

Source of data	Number of studies	Evolutionary rate (*darwins*)	Time interval studied
Artificial selection experiments	8	12,000 to 200,000	1.5 to 10 years
Colonization	104	0 to 79,700	70 to 300 years
Mammal fossils since the last ice age	46	0.11 to 32.0	1,000 to 10,000 years
Fossil invertebrates	135	0 to 3.7	0.3 to 350 million years
Fossil vertebrates	228	0 to 26.2	0.008 to 98 million years

Data from Gingerich (1983).

span tens to hundreds of thousands of generations, and because the *average* rate of change is always less than the *fastest* rate, short-term rapid variations are likely to be missed. Unlike studies of colonization and artificial selection where changes can be monitored from one generation to the next and true rates of change, not just their average, can be measured, studies of the fossil record almost always underestimate the fastest rates of evolution of a lineage.

Analysis of the rich fossil record of the foraminiferan lineage from *Globorotalia inflata* to *Globorotalia conoidea* demonstrates how coarseness of sampling can lead to a gross underestimate of the rate of evolution. Figure 3.13A shows that members of the lineage changed in size from 8.3 million years ago to about 10,000 years ago at an average (and very slow) rate of about 0.015 *d*. The specimens on which the study is based were collected from a deep-sea drill core which when sampled at close intervals throughout its length reveals that instead of a simple slow increase in size (Figure 3.13A) the lineage actually underwent many rapid size-changes (Figure 3.13B). The fastest average rate is 4.6 *d*, some 300 times faster than the overall average, and even faster rates would probably be revealed were the core sampled still more finely!

But closer-spaced sampling does not always reveal such a dramatic difference. For example, in P. R. Sheldon's (1987) study of the change in the number of riblike **thoracic** segments in **trilobites** collected from Welsh rocks of the **Ordovician Period** of geologic time (extending from about 495 to 440 million year ago), evolutionary rates measured by fine sampling (Figure 3.14, below) proved to be about the same as those determined by much coarser sampling (Figure 3.14, above), especially for trilobites of the genus *Nobiliasaphus* (shown at the far right of the figure). Yet even in this study, closer-spaced samples reveal differences from the overall average, such as for species of *Ogygiocarella* (the lineage to the left of *Nobiliasaphus*) for which the fastest rate measured by fine sampling is almost ten times more rapid than shown by the coarse.

The striking, though artifactual contrast evident in Table 3.2 between the seeming slowness of evolution measured over geological time and its rapidity over biological time may in part also be due to the types of traits available for paleonto-

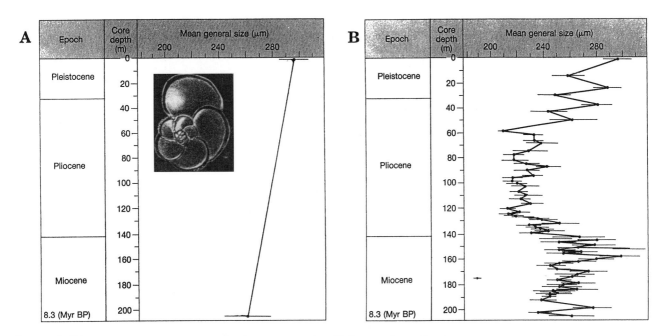

FIGURE 3.13 The detected rate of evolution increases greatly with increased sampling in the four-species lineage of foraminifera extending from Miocene *Globorotalia conoidea* to Pleistocene *Globorotalia inflata*. (A) On the basis of only two widely spaced samples, one might mistakenly conclude that the evolutionary size increase of this lineage happened gradually. (B) However, sampling at finer intervals reveals a much more complex evolutionary path and high rates of change, the fastest about 300 times faster than the overall average shown in (A) (Modified after Buzas *et al.*, 1987, and Ridley, 1996, and based on data from Malmgren and Kennett, 1981.)

logic study (Ridley, 1996). In particular, because paleontologic time scales are so long and the fossil record so incomplete, rapidly evolving characters simply change too fast to leave a recognizable record so paleontologists have no choice but to focus on slowly evolving traits.

How Fast Can Meristic Traits Evolve?

So far we have considered only the evolution of continuous characters, but meristic traits also show many interesting evolutionary changes. Perhaps the best known example is the evolutionary sequence from five-toed ancestral horses to the single-toed hoofs of their living descendants. This transition spanned about 50 million years and appears to have happened in spurts rather than as a slow steady progression brought about by continual selection (MacFadden, 1992). How fast might it have happened had there been steady selection for reduction in the number of toes?

R. Lande (1978) developed a quantitative model to answer this question. By choosing a very low heritability for toe number ($h^2 = 0.1$, lower even than the low value for the number of offspring per litter in mice; Table 3.1) and a low selective advantage for having fewer toes, he biased his model *against* rapid evolutionary change. Even so, he found that under sustained selection the evolution from five toes to one could happen in as short a time as only a million years! This conclusion is of course only as firm as the assumptions embodied in Lande's model, but it does

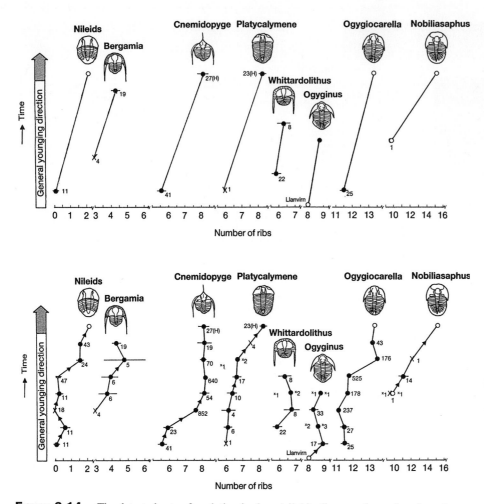

FIGURE 3.14 The detected rate of evolution in these trilobite lineages (spanning about 3 million years) is about the same whether fossil sampling is coarse (above) or at much finer intervals (below), a similarity especially evident for the genus *Nobiliasaphus* (far right). (Modified after Sheldon, 1987.)

appear that under appropriate conditions meristic characters can evolve very fast, much faster than rates measured in the fossil record (MacFadden, 1992).

Evolution Usually Proceeds More Slowly Than Its Maximal Rate

Though factors such as coarseness of sampling cause evolutionary rates measured in the fossil record to appear much slower than those observed over years or a few centuries, it nevertheless is likely that the fastest rates measured in laboratory experiments (Table 3.2) are much faster than the evolution of most lineages over most of their histories. The selection pressures applied in artificial selection experiments are often relentless and ferociously severe. In nature, evolution almost always happens much more slowly than the high rates to which it can be forced in the laboratory or shown in quantitative models able to achieve—evolution almost always operates much more slowly than it actually can!

Evolution Can Be Seen to Operate on Very Short Time Scales

Because in most species evolution proceeds too slowly to be measured over one or even several human lifetimes, some have argued that evolutionary theory cannot be tested the same way other scientific theories are. There are, however, good examples where evolutionary change can be seen happening even over only a few years, such as in breeding experiments on the fruit fly, *Drosophila melanogaster.* The results of one experiment, in which flies were selected for breeding based on the number of bristles on their abdomens, are shown in Figure 3.15. Dramatic changes were seen after less than three dozen generations each lasting about two weeks, a total time of less than two years!

A even more telling example comes from studies of the potency of certain **viruses** that is due in part to the remarkable speed of their evolution. For instance, the **influenza** virus evolves so quickly that new vaccines must be developed yearly. And **HIV (human immunodeficiency virus)**, the viral cause of **AIDS (acquired immunodeficiency syndrome)**, evolves so fast that evolutionary differences can be detected in samples taken from the same person just a few years apart. Indeed, the **protein** coats of HIV and influenza viruses evolve faster than our bodies can develop **antibodies** to fight them, so the rapidity of their evolution is a major factor that makes it so it difficult to devise effective treatments.

THE GENETIC BASIS OF MORPHOLOGICAL CHANGE

Let us now turn to another, perhaps even more fascinating aspect of evolutionary change, namely, how does evolutionary innovation result in the origin of a brand new species?

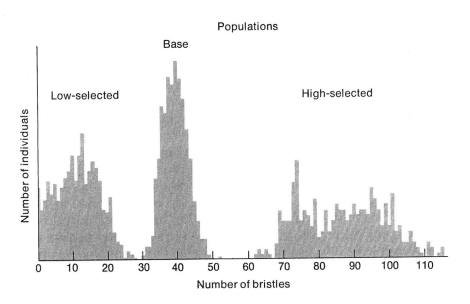

FIGURE 3.15 In artificial selection experiments, evolutionary change can induced to happen in a very short time. Selective breeding of fruit flies (*Drosophila melanogaster*) having a low or high number of abdominal bristles from a base (starting) population results in dramatic evolution of the average number of bristles in populations of offspring produced in fewer than three dozen generations (a period of less than two years). (Modified after Futuyma, 1986, and based on data from Clayton and Robertson, 1957, and Falconer, 1981.)

Figure 3.8 compares the two main models for the evolution of a biologic lineage, phyletic gradualism and punctuated equilibrium. Each of the diagrams in the figure shows a single species diverging into two species by a single speciation event. In the phyletic gradualism model (Figure 3.8, left), slow morphological change happens throughout the evolution of the lineage whereas in the punctuated equilibrium model (Figure 3.8, right) morphological change is concentrated and rapid when species originate. Since morphological changes of this type are determined by underlying genetic changes, two crucial questions need to be addressed: What is the nature of the genetic changes responsible for the differences between species? And must there be a continuum of changes between species or can the changes be abrupt and discontinuous? We are still a long way from knowing complete answers to these questions, largely because we know relatively little about the genetic and developmental bases for the differences between species. But as we now shall see, important progress is being made.

Small Genetic Changes Can Have Major Morphological Effects

Though members of a species usually show a gradation of variability in morphological traits, rare individuals that have very different morphologies are occasionally present. For example, more than a century ago William Bateson (1894) discovered an individual moth of the species *Zygoena filipena* which in place of one of its legs had an additional wing. The moth was a mutant, a so-called freak of nature, in which one part of its body was replaced by another, called a **homeotic mutant**. We now know that this type of mutation can be caused by a very small genetic change, the loss of function of but a single gene.

Homeotic mutants have been found in many different species, even humans, but are especially well studied in the fruit fly, *Drosophila melanogaster* (De Robertis, 1995). One of the first to be studied is a mutation that causes flight-stabilizing structures (**halteres**) on the third thoracic segment of a fruit fly to be replaced by wings so that the third segment is transformed to look like an extra copy of the second thoracic segment. This homeotic transformation results when a gene called **Ultrabithorax** (*Ubx*) loses function. The *Ubx* gene is but one of a series of similar genes, members of the **HOM/Hox** gene family, present in animals where they play a crucial role in determining the structural makeup of body parts along the length of the organism (Carroll, 1995; De Robertis, 1995).

Some Major Morphological Changes Result from Simple Genetic Changes

Not only is discovery of the genetic basis of naturally occurring homeotic mutations one of the most spectacular successes of modern genetics, it also provides a key to understanding the type of genetic change responsible for at least some differences between species (Shubin *et al.*, 1997). In various kinds of **crustaceans** such as crabs and shrimp, for example, the number and types of appendages on body segments tend to vary from one segment to the next. M. Averof and N. H. Patel (1997) have shown that certain of these variations depend on whether or not two particular *HOM/Hox* genes, *Ubx* and *Abd-A*, are genetically "turned on" when a segment forms—feeding appendages (**maxillipeds**) are made if the genes are expressed, but if they are "turned off" a different type of limb develops. A simple on-off switch governing expression of but two genes evidently played a major role in the evolution of the many different kinds of crustaceans.

Changes in the expression of *HOM/Hox* genes may also have been important in the evolution of insects. It appears that the genetic "default" condition for each segment of an insect is to develop wings, and that over the course of evolution wing development came to be suppressed except in thoracic segments or a subset of them (Carroll *et al.*, 1995). This idea gains support from the fossil record in the form of the beautifully preserved 270 million year-old mayfly **nymph** shown in Figure 3.16 which has small winglike structures (probably used for swimming, not flying) down the length of its entire body.

Genetic Changes Can Give Rise to Neomorphs

None of the gene-governed evolutionary transformations just discussed involve the origin of a completely new structure. Instead, they show how one type of preexisting structure, such as a haltere or a wing, may have been replaced by another, in some sense a fairly simple change. But how did halteres and wings originate in the first place? How do truly new complex structures come into being?

Much remains unknown about the genetic changes that underlie the origin of new structures, neomorphs, though it is likely the changes are many and complicated. We are, however, beginning to see the first glimpses of how some complex evolutionary transformations happened (Sordino *et al.*, 1995; Nelson and Tabin, 1995; Shubin *et al.*, 1997). Perhaps the one we know most about is the evolutionary sequence from the fins of fish to the limbs of terrestrial vertebrates, **tetrapods**. Structures intermediate between fish fins and tetrapod limbs are present in both of the two types of living **lobe-finned fish**, the **lungfish** and the **coelacanth**. Most living fish—such as a goldfish, for example—have fins supported by very fine rays (and for this reason are called **ray-finned** fish), whereas the fins of lobe-finned fishes are stout structures that like tetrapod limbs consist of muscles and bone. And as shown in Figure 3. 17, the pattern of major bones in the fins of certain fossil lobe-finned fish is very similar to that of the same set of bones in tetrapod limbs.

Early studies of this transformation proposed that the long axis of the fins of lobe-finned fish (Figure 3.18A) correspond to the long axis of tetrapod appendages (Figure 3.18B) and that minor bony elements in the fins came to be modified to tetrapod digits (fingers and toes). Later studies proposed that during evolution to tetrapod limbs, the primitive lobe-fin came to be bent over in such a way that tetrapod fingers and toes were actually brand new structures developed on the outer side of the bent fin (Figure 3.18C), a view now supported by gene expression experiments. In fish, a particular *HOM/Hox* gene known as **Hoxd-11** is expressed in the rear (posterior) part of the bud that develops into a fin (Figure 3.18D). In tetrapods the same gene is also first expressed in the posterior part of developing limb-buds but later expressed also at the tip of the buds where the digits eventually form (Figure 3.18E). Up to now, the gene study has been carried out only on a ray-finned fish (Sordino *et al.*, 1995) so it will be important to see whether gene expression in lobe-finned fish follows the same pattern.

Even Changes in Continuous Characters May Result from Simple Genetic Changes

Though the homeotic transformation of one structure to another and the origin of entirely new structures are among the most interesting types of evolutionary change, most morphological differences between any two closely related species are in the relative sizes and shapes of structures common to both. As we have seen, continuous traits usually show a gradation of form among members of a species, and for this reason it is often assumed that such traits evolve gradually and smoothly from one species to another. It is not obvious, however, that this must always be true for it is at least plausible that even continuous characters may evolve discontinuously.

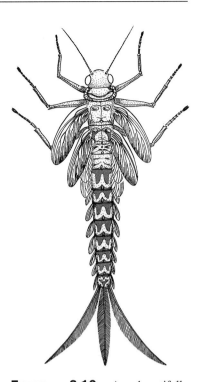

FIGURE 3.16 A beautifully preserved fossil mayfly nymph (juvenile) from Permian rock strata of Oklahoma having paired winglike structures on each body segment. Recent studies seem to show that each body segment of an insect will develop wings, as shown in this fossil, unless the genetic instructions are switched off. The ability to turn off wing-making genes in specific body segments appears to have been an important innovation in insect evolution. (Modified after Kukalova-Peck, 1978.)

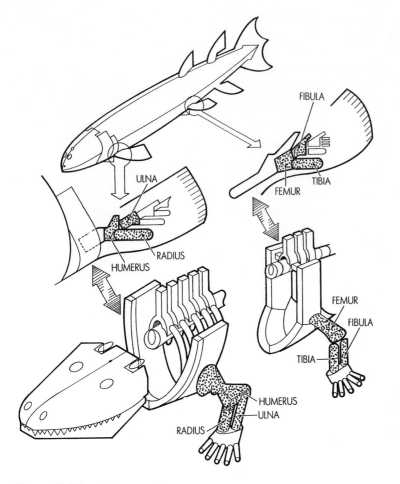

FIGURE 3.17 Evidence of the evolutionary transition from fish to terrestrial vertebrates is shown by the similar bones (denoted by stippling) in the fins of some lobe-finned fish (such as the Devonian-age *Eusthenopteron* shown at the upper left) and the limbs of tetrapods (represented by an early-evolved amphibian). (Modified after Radinsky, 1987.)

A well known genetic disorder in humans is **Down's syndrome**, manifest by malformation of the skull and face and caused by the presence of three, rather than the normal two, copies of chromosome 21. The extra copy results in over-expression of the chromosomes' genes including one in particular, *Ets2*, which plays an important role in the formation of cartilage including skull precursor cells. As shown by S. H. Sumarsono *et al.* (1996), over-expression of the same *Ets2* gene in mice produces similar results (Figure 3.19). This discovery from modern genetics is important to evolution because it illustrates that dramatic changes of morphology can take place without invention of new genes or modification of existing ones— minor alterations of when genes are switched on or off, or simple increases or decreases in the levels of expression of even a single gene, can induce large morphological changes.

Though the *Ets2* example shows that change of a single gene might lead to the discontinuous evolution of continuous traits, this would happen only very rarely. Dramatic changes in morphology, such as those from over-expression of the *Ets2* gene, are likely to be seriously deleterious. As a consequence, changes of this type tend to be eliminated rapidly by natural selection so it is much more common

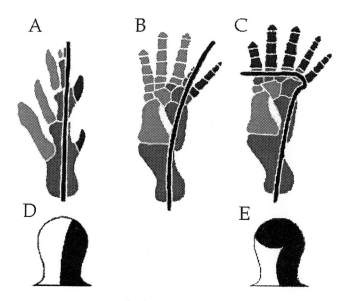

FIGURE 3.18 Schematic drawings summarizing two models for the origin of tetrapod limbs from fish fins. (A) The skeleton of the fin of a lobe-finned fish (with a narrow rod superimposed to denote the long axis of the fin). (B) An early hypothesis for the derivation of the tetrapod limb viewed its long axis (denoted here by the slightly curved superimposed rod) as corresponding to the long axis of the lobe-fin and the limb's digits (fingers and toes) as modified fin bones. (C) The current hypothesis is that during evolution the long axis of the lobe-fin came to be bent (denoted by the hook-shape of the superimposed rod) and the digits arose as new structures, neomorphs protruding from the outside edge of the deflected bony axis. (D) Diagram showing (in black) the rear (posterior) position of the expression of the *Hoxd-11* gene in the developing fin-bud of a ray-finned fish (in this experiment, a zebra fish). (E) In tetrapods, the expression of *Hoxd-11* (region shown in black) happens first in the posterior part of a limb-bud then also later where digits eventually develop (shown here in a diagram of the limb-bud of a mouse). Comparison of the pattern of gene expression in fish and mouse supports the current hypothesis that tetrapod limbs evolved by a change in orientation of the long axis of the lobe-fin. (Modified afterNelson and Tabin, 1995.)

for new species to arise from the cumulative effects of small changes in many genes. And even a rare evolutionarily successful change in expression of a single gene would doubtless lead to a cascade of subsequent changes in the expression of a whole host of other genes as the organism's new form is honed by selection.

ORIGIN OF *HOMO SAPIENS* REVISITED

The known hominid fossil record is incomplete, especially for the earliest evolved forms. Nevertheless, numerous finds have been made (Figure 3.20), and though the exact number of genera and species is open to debate it is striking that for much of hominid history a few or several different kinds coexisted whereas at present the group is represented by but a single surviving member—us!

This chapter began with an analysis of how Dubois' discovery of Haeckel's hypothetical missing link contributed to understanding human origins. In light of the material reviewed since, let us now return to three questions about the origin of humans and humanness: What sequence of major evolutionary innovations happened since the last common ancestor of humans and chimpanzees? What was

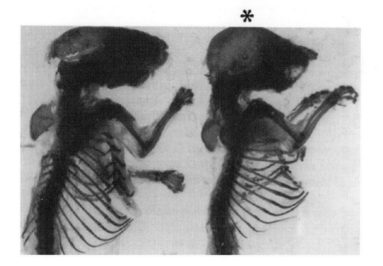

Figure 3.19 Normal (left) and experimentally altered (right) mouse skeletons showing that the abnormalities of Down's syndrome can be mimicked merely by causing an increase in the level of expression of a single gene. Abnormalities seen typically in humans having Down's syndrome, such as a domed forehead (*) and shortened face, are induced by artificially increasing the level of expression of the *Ets2* gene. (Modified after Sumarsono *et al.*, 1996.)

the rate of brain size increase in the transition from *Homo erectus to Homo sapiens*? And what does the fossil record reveal about the origin of language, the characteristic Haeckel supposed diagnostic of modern humans?

The Sequence of Some Major Human Evolutionary Innovations

To assess the order in which major innovations arose in the evolution of *Homo sapiens* from our last common ancestor with chimpanzees, I will focus on the two hominid fossils that are most famous (largely because they are the most complete), "Lucy" and the "Nariokotome boy." Lucy, unearthed from 3.2 million-year-old rock strata at Hadar in Ethiopia, East Africa, is the most complete specimen of *Australopithecus afarensis* yet discovered. The Nariokotome boy is a remarkably well-preserved specimen of *Homo erectus*, found in 1.5-million-year-old strata of northern Kenya, south of Ethiopia (Walker and Shipman, 1996).

Shown in Figure 3.21 is the sequence of evolutionary events these fossils reveal, beginning with the loss of a tail (a character shared by chimpanzees and all hominids) followed by the sequential development of bipedality and a low broad pelvis (shared by Lucy, the Nariokotome boy, and *Homo sapiens*); a barrel chest and long legs (shared by the two species of *Homo*); and a chin, large brain, and language (characters evidently restricted to *Homo sapiens*, though the precise placement of language on the hominid branch of the phylogenetic tree remains somewhat uncertain). It is truly notable that on the basis of just two well preserved fossil skeletons, Lucy and the Nariokotome boy, it is possible to construct this fairly complete picture of the evolutionary sequence from our last common ancestor with chimpanzees to humans today. Other fossils confirm the sequence shown and give additional insight about the nature of the evolutionary transition (see, for example, Conroy, 1997).

One particularly spectacular piece of evidence corroborating the early origin of an upright stance and bipedality comes from a trail of unmistakably hominid footprints preserved in a bed of volcanic ash (**tuff**) at Laetoli in Tanzania, East Africa. Because only fossils less than about 50,000 years old can be dated directly (by **carbon-14 dating**, as discussed in Chapter 2), the precise age of older fossil hominid finds is often problematic. But because volcanic ash is of **igneous** origin, derived from once molten rock, tuffs older than 50,000 years can be dated (by isotopic methods explained in Chapter 2), and the footprint-bearing ones at Laetoli have an age of about 3.5 million years or slightly older. The footprints are presumed to have been made by Lucy's species, *Australopithecus afarensis*, since bones of this hominid are also present at Laetoli, so the site confirms that Lucy and her kin walked upright and that this form of locomotion evolved well before most other major traits that distinguished us from chimpanzees and the chimpanzee-hominid ancestor.

Despite the notable advances made by modern genetics toward understanding the mechanisms underlying morphological change, we still have only rudimentary knowledge of the genetic changes that led to modern humans, including those

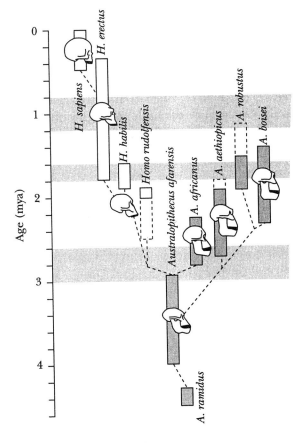

FIGURE 3.20 An evolutionary tree of hominid fossils. Hominids diversified rapidly beginning about 3 million years ago (mya) and for much of subsequent geologic time were represented by two or more coexisting species. For example, ancestral *Homo erectus* and its descendant, *Homo sapiens,* coexisted from about 0.5 to 0.4 mya, an overlap in time which makes it difficult to determine exactly when the transition happened. (Modified after Conroy, 1997.)

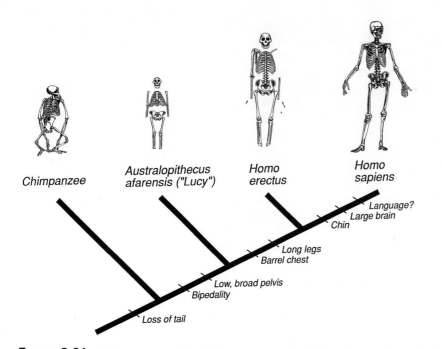

Chimpanzee

Australopithecus
afarensis ("Lucy")

Homo
erectus

Homo
sapiens

Language?
Large brain
Chin

Long legs
Barrel chest

Low, broad pelvis
Bipedality

Loss of tail

FIGURE 3.21   The sequence of evolutionary events that led to humans from the last common ancestor of chimpanzee and *Homo sapiens* can be ordered with the help of just two well preserved intermediate-aged hominid fossils, "Lucy" *(Australopithecus afarensis)* and the "Nariokotome boy" *(Homo erectus)*. The drawings of the skeletons show the relative sizes of the actual specimens. (Modified after Conroy, 1997, and Jurmain *et al.*, 1997.)

responsible for the evolution of our large brains and the capacity for language. But as we will now see, the fossil record does provide interesting and important insight about the evolutionary history of these two evidently uniquely human attributes.

How Fast Was the Rate of Increase in Human Brain Size?

Earlier we examined the roles natural selection and heritability play in driving evolution and the rates at which evolution proceeds. How does the rate of evolution of the remarkably large human brain compare with evolutionary rates of other characters in other species?

The cells and tissue of brains are not themselves geologically preservable, so brain size cannot be measured directly in fossils. But a brain fits quite snugly into its skull so the volume within a skull, the **cranial capacity**, is a good proxy for brain size (even though it overestimates the actual volume by a small amount). Estimates of the cranial capacity (in cubic centimeters, cc) of chimpanzees and some major hominid species are listed in Table 3.3. Because relatively few skulls of the fossil hominids are known and most of these are damaged it often is difficult to measure accurately their cranial capacities. This problem is confounded by uncertainty among paleoanthropologists about which fossils belong to which species or even the number of true species discovered. The cranial capacities listed in Table 3.3 should therefore be viewed as rough approximations only.

As shown in Table 3.3, the cranial capacity of modern *Homo sapiens* is about one-and-a-half times larger than the average cranial capacity of its evolutionary ancestor, *Homo erectus*. However, this 50% rise may not be an accurate index of the actual size increase. *Homo erectus* crania have varied volumes—those from Zhoukoudian, China, are on average 1,030 cc, whereas African specimens average only 870 cc (Conroy, 1997)— and there are not enough specimens known to calculate a firm overall average for the species as can be done for *Homo sapiens*. Given that *Homo erectus* populations vary and there is no way to know from which population *Homo sapiens* was derived (or even if specimens of that particular population have been discovered), the precise amount of cranial capacity increase between *Homo erectus* and us is unknown but there is no doubt that brains came to be larger, evidently by a factor between 35% and 60%.

How fast did this size increase happen and how does its rate compare with rates of change in other species? This seems a simple problem—we have a rough measure of the size increase, so to figure the rate of change all we need is an estimate of the time spanned. The transition time, however, is not easy to estimate accurately because the fossil record shows that for a time the two species actually coexisted (Figure 3.20). Evidently, *Homo erectus* did not become extinct until after the time of origin of *Homo sapiens* roughly 500,000 to 400,000 years ago.

If the emergence of *Homo sapiens* from *Homo erectus* took a million years (probably an overestimate), the average rate of cranial capacity increase was slow, a meager 0.4 *darwins*. If the transition took 100,000 years, the rate was 4.0 *d*; and if 10,000 years (likely an underestimate), 40 *d*, a rate still notably slow, more than three orders of magnitude (1,000 times) slower than the fastest rates measured in artificial selection experiments or for living species colonizing new habitats (Table 3.2). And even these fastest measured rates are probably slight overestimates because the earliest *Homo sapiens*, known as **archaic *Homo sapiens*** (and dating from about 400,000 to 125,000 years ago) had smaller cranial capacities than modern humans, about 1,250 cc compared to 1,390 cc (Table 3.3). So not all of the size increase happened in the transition from *Homo erectus* to modern *Homo*

TABLE 3.3

Cranial capacities of chimpanzees and some hominid species.

Species	Average volume (cc)	Range of volumes* (cc)
Chimpanzees	395	320 to 470
Australopithecus afarensis	415	350? to 500?
Australopithecus boisei	465	330 to 595
Homo habilis	640	430 to 850
Homo erectus	935	645 to 1,230
Homo sapiens (archaic)	1,250	1,000 to 1,400
Homo sapiens (modern)	1,390	1,170 to 1,610

Volumes listed are approximate—they are not adjusted for differing body sizes of the various species and ignore differences in cranial capacity between males and females and variation between different populations of individual species. Data from Conroy (1997) and Jurmain *et al.* (1997).

* 95% of individuals in species

sapiens for which the best estimate of the rate of change is between 3.0 and 30 *d*. Clearly, the emergence of the largest-brained hominids, *Homo sapiens*, was not an especially fast transition.

(Interestingly, evidence shows that Neanderthal man, not listed in Table 3.3, had an average cranial capacity of 1,400 cc, slightly larger than the average for modern *Homo sapiens*, so we living humans may not actually be the largest-brained hominids to have existed.)

When Did Language First Arise?

In the 1800s, Ernst Haeckel suggested language to be a defining characteristic of modern humans and accordingly named his hypothetical missing link the ape-man without speech, *Pithecanthropus alalus*. But when Haeckel's missing link was later unearthed by Eugène Dubois there was no way to tell whether the hominid did or did not possess language so Dubois avoided the issue by naming his species *erectus* for its ability to walk upright, a characteristic deduced directly from the fossils.

Now, more than a century later, what does the fossil record tell us about the time of origin of the capacity for language? At first glance this may seem unanswerable—after all, as a behavior carried out by living, breathing people, language can hardly be expected to have left its mark in the rock record. Yet it is conceivable the fossil record might harbor some *indirect* evidence of language, for example of a language-requiring activity that leaves a physical trace. One possibility are hand-tools such as those associated with *Homo habilis* dating from about 2.5 million years ago (Potts, 1993) which would give strong proof of language if tool-making and tool-using could not happen without complicated verbal communication. However, chimpanzees and even certain kinds of birds fashion tools but do not posses language, so tool-making may have come long before language. And though even fragments preserved of a written language would be irrefutable evidence, speech may well have developed well before writing.

A more promising possibility is close inspection of skeletal remains of early hominids. Different parts of a human brain govern different specific functions, such as **Broca's area** which plays a critical role in the production of language. Significantly, Broca's area leaves an imprint of its shape and surface texture on the

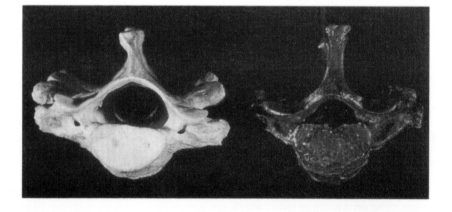

FIGURE 3.22 Comparison of neck vertebrae from a modern human (left) and the Nariokotome boy (right) showing that the hole (vertebral foramen) through which the spinal cord passes is much larger in *Homo sapiens* than *Homo erectus*. (From Walker, 1993, republished with permission.)

inside of a skull. Because Dubois' *Pithecanthropus erectus* (now *Homo erectus*) has an impression of Broca's area, it may have been able to speak (though this evidence was unknown to Dubois since it was not until much later that sediment was removed from the fossil skull cap to expose its inside surface). If so, the capacity for language would appear to extend to at least 1.8 million years ago, the age of the oldest known *Homo erectus*, and since imprints of Broca's area are present also on nearly 2-million-year-old skull caps of *Homo habilis*, a hominid known earliest from deposits 2.4 million years in age, some feel that our lineage has possessed at least a rudimentary form of language for over 2 million years.

Others, however, point to recent discoveries that suggest speech may have originated very much later, *after* the origin of *Homo sapiens* some 500,000 to 400,000 years ago (Figure 3.20). The thickness of the **spinal cord** in humans is much larger than in chimpanzees, a difference easy to see from the size of the hole (the **vertebral foramen**) through which it passes down the center of each **vertebra** of the backbone. The nerve-packed cable leaving the skull in humans is thicker and carries many more nerve fibers than in the great apes. Interestingly, the size of the vertebral foramen of the Nariokotome boy, the most complete known specimen of *Homo erectus*, is chimplike rather than humanlike (Figure 3.22). A. Walker and P. Shipman (1996) theorize that the extra nerves in the human spinal cord control muscles around the diaphragm without which speech would be impossible, and on this basis suggest the narrow vertebral foramen in *Homo erectus* shows it could not speak.

How can we reconcile the two lines of evidence, the narrow vertebral foramen that evidently shows *Homo erectus* unable to speak and the presence of Broca's area, the brain center associated with motor control of speech, both in *Homo erectus* and even older *Homo habilis*? The answer comes from recent studies

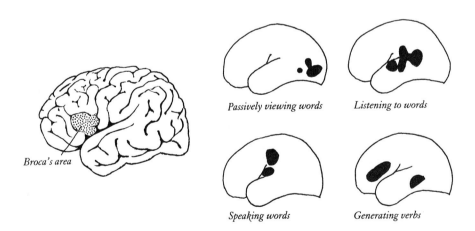

Broca's area

Passively viewing words   *Listening to words*

Speaking words   *Generating verbs*

FIGURE 3.23 Outline drawings of human brains (oriented with their fronts at the left) showing regions used to produce and comprehend language. Studies of inside surfaces of fossil skulls show that both *Homo erectus* and *Homo habilis* possessed Broca's area (denoted in the drawing at the left), a brain center associated with the motor control of speech, which suggests these species may have been capable of speech. However, studies of brain activity (measured by positron emission tomography, PET) in living humans demonstrate that many other regions (areas shown in black) are also involved in producing and comprehending language and that for some such tasks Broca's area is not used. Despite the presence of Broca's area in *Homo erectus* and *Homo habilis*, neither would have been able to speak if they lacked these additional parts of the brain. (Modified after Walker and Shipman, 1996.)

in the field of **neurophysiology** that give vastly improved insight into the way the human brain works. In particular, we now know that Broca's area alone is not sufficient to produce speech, that many other regions in addition are required, and that Broca's area is not even needed for some components of language (Figure 3.23). But since many of the other regions involved in speech do not leave telltale imprints on skulls it is not possible to use skull caps to decide whether a fossil hominid had all the necessary components.

So the two lines of evidence are not in conflict—Brocas's area, present in *Homo habilis* and passed on to *Homo erectus*, evolved early. But a thickened spinal chord and capacity for language evolved later, evidently after the divergence of our own species, *Homo sapiens*, from our *Homo erectus* ancestors. Language is probably younger than 500,000 years, perhaps much younger—as Walker and Shipman (1996) point out, it would appear that Dubois' fossil was indeed Haeckel's speechless missing link!

CONCLUSIONS

The hunt for missing links in the fossil record—dating from the 1800s when it was spurred by Darwin's insight, proposed formally by Haeckel, and pursued with vigor by Dubois—is bound to prove practically fruitless: Ancestors cannot be identified with clear cut certainty in the fossil record. Yet we have seen that the fossil record is a rich source of scientific data by which to trace the course of evolution. So, if we are willing to replace the Haeckelian definition of "missing link" found in many dictionaries (Box 3.1) with a sharpened up-to-date version (Box 3.2), it becomes apparent that the fossil record is actually replete with missing links and that these provide powerful evidence of the nature and path of evolutionary change. Now, for the first time, breakthroughs in developmental biology and genetics make it possible to begin to fathom the mechanisms that underlie the morphological changes shown in the fossil record. By combining the methods, facts, and concepts of paleontology, evolutionary biology, and developmental genetics, we are poised on the threshold of firm understanding of our origins.

Box 3.1   **Traditional Definition**

missing link *n.* 1. (usually preceded by *the*) an extinct animal postulated to bridge the evolutionary gap between the anthropoid apes and human beings.

• Named *Pithecanthropus alalus* (ape-man without speech) by Haeckel in 1868, and maniacally sought by Dubois in the Dutch East Indies (now Indonesia) in the last years of the nineteenth century.

Box 3.2   **New Definition**

missing link *n.* 1. *Not standard* (and never to be preceded by *the*). A fossil:

• Whose **morphology** may be used to help determine the order of appearance of evolutionary innovations;

• Whose **age** may be used to place a minimum estimate on the time of origin of evolutionary innovations.

Acknowledgments

I thank Bill Schopf, Michele Nishiguchi, and Marilyn McCoy for comments on this essay, Harry Jerison for help with the hominid cranial capacity data, and Richard Mantonya for assistance with the figures.

References

Averof, M. and Patel, N.H. 1997. Crustacean appendage evolution associated with changes in Hox gene expression. *Nature* **388**: 682-686.

Bateson, W. 1894. *Materials for the Study of Variation Treated with Especial Regard to Discontinuity in the Origin of Species* (London: Macmillan), 598 pp.

Buzas, M.A., Douglass, R.C., Smith, C.C. 1987. Kingdom Protista. In: R.S. Boardman, A.H. Cheetham and A.J. Rowell (Eds.), *Fossil Invertebrates* (Palo Alto, CA: Blackwell), pp.67-106.

Carroll, S.B., Weatherbee, S.D. and Langeland, J.A. 1995. Homeotic genes and the regulation and evolution of insect wing number. *Nature* **375**: 58-61.

Carroll, S.B. 1995. Homeotic genes and the evolution of arthropods and chordates. *Nature* **376**: 479-485.

Cavalli-Sforza, L.L. and Bodmer, W.F. 1971. *The Genetics of Human Populations* (San Francisco: Freeman), 965 pp.

Clayton, G.A. and Robertson, A. 1957. An experimental check on quantitative genetical theory. II. The long-term effects of selection. *J. Genetics* **55**: 152-170.

Colbert, E.H. 1980. *Evolution of the Vertebrates* (New York: Wiley), 510 pp.

Conroy, G.C, 1997. *Reconstructing Human Origins* (New York: Norton), 557 pp.

Darwin, C. 1859. *The Origin of Species*, 1st Edition (London: John Murray), 502 pp.

De Robertis, E.M. 1995. Homeotic genes and the evolution of body plans. In: C.R. Marshall and J.W. Schopf (Eds.), *Evolution and the Molecular Revolution* (Boston: Jones and Bartlett), pp.109-124.

Devlin, B, Daniels, M. and Roeder, K. 1997. The heritability of IQ. *Nature* **388**: 468-471.

Dodson, P. 1990. Counting dinosaurs: How many kinds were there? *Proc. Natl. Acad. Sci. USA* **87**:7608-7612.

Eldredge, N. and Gould, S.J. 1972. Punctuated equilibrium: An alternative to phyletic gradualism. In: T.J.M. Schopf (Ed.), *Models in Paleobiology* (San Francisco: Freeman, Cooper and Co.), pp.82-115.

Endler, J.A.. 1986. *Natural Selection in the Wild* (Princeton, NJ: Princeton University Press), 336 pp.

Erwin, D.H. and Anstey, R.L. 1995. *New Approaches to Speciation in the Fossil Record* (New York: Columbia University Press), 342 pp.

Falconer, D.S. 1981. *Introduction to Quantitative Genetics*, 2nd Edition (London: Longman), 340 pp.

Foote, M. 1996. On the probability of ancestors in the fossil record. *Paleobiology* **22**: 141-151.

Futuyma, D.J. 1986. *Evolutionary Biology* (Sunderland, PA: Sinauer), 600 pp.

Gingerich, P.D. 1974. Stratigraphic record of Early Eocene *Hyopsodus* and the geometry of mammalian phylogeny. *Nature* **248**: 107-109.

Gingerich, P.D. 1983. Rates of evolution: effects of time and temporal scaling. *Science* **222**: 159-161.

Gould, S.J. 1980. *The Panda's Thumb* (New York: Norton), 343 pp.

Jurmain, R., Nelson, H., Kilgore, L. and Trevathan, W. 1997. *Introduction to Physical Anthropology*, 7th Edition (Belmont, WA: Wadsworth), 600 pp.

Kukalova-Peck, J. 1978. Origin and evolution of insect wings and their relation to metamorphosis, documented by the fossil record. *J. Morphology* **156**: 53-126.

Lande, R. 1978. Evolutionary mechanisms of limb loss in tetrapods. *Evolution* **32**: 73-92.

MacFadden, B.J. 1992. *Fossil Horses* (Cambridge, UK: Cambridge University Press), 369 pp.

Malmgren, B.A. and Kennett, J.P. 1981. Phyletic gradualism in a Late Cenozoic planktonic foraminiferal lineage; DSDP Site 284, southwest Pacific. *Paleobiology* 7: 230-240.

Malmgren, B.A., Berggren, W.A. and Lohmann, G.P. 1983. Evidence for punctuated gradualism in the Late Neogene *Globorotalia tumida* lineage of planktonic foraminifera. *Paleobiology* **9**: 377-389.

Marshall, C.R. 1995. Darwinism in an age of molecular evolution. In: C.R. Marshall and J.W. Schopf (Eds.), *Evolution and the Molecular Revolution* (Boston: Jones and Bartlett), pp.1-30.

Martin, R.D. 1993. Primate origins: Plugging the gaps. *Nature* **363**: 223-234.

Mayr, E. 1991. *One long Argument* (Cambridge, MA: Harvard University Press), 195 pp.

Nelson, C.E. and Tabin, C. 1995. Footnote on limb evolution. *Nature* **375**: 630-631.

Potts, R. 1993. Archeological interpretations of early hominid behavior and ecology. In: D.T. Rasmussen, D.T. (Ed.), *The Origin and Evolution of Humans and Humanness* (Boston: Jones and Bartlett), pp.49-74.

Radinsky, L.B. 1987. *The Evolution of Vertebrate Design* (Chicago: University of Chicago Press), 188 pp.

Ridley, M. 1996. *Evolution*, 2nd Edition (Boston: Blackwell), 752 pp.

Sheldon, P.R. 1987. Parallel gradualistic evolution of Ordovician trilobites. *Nature* **330**: 561-563.

Shubin, N., Tabin, C. and Carroll, S. 1997. Fossils, genes and the evolution of animal limbs. *Nature* **388**: 639-648.

Sordino, P., van der Hoeven, F. and Duboule, D. 1995. *Hox* gene expression in teleost fins and the origin of vertebrate digits. *Nature* **375**: 678-681.

Sumarsono, S.H., Wilson, T.J., Tymms, M.J., Venter, D.J., Corrick, C.M., Kola, R., Lahoud, M.H., Papas, T.S., Seth, A. and Kola, I. 1996. Down's syndrome-like skeletal abnormalities in *Ets2* transgenic mice. *Nature* **379**: 534-537.

Tobias, P.V. 1992. Major events in the history of mankind. In: J.W. Schopf (Ed.), *Major Events in the History of Life* (Boston: Jones and Bartlett), pp.141-175.

Walker, A. 1993. The origin of the genus *Homo*. In: D.T. Rasmussen, D.T. (Ed.), *The Origin and Evolution of Humans and Humanness* (Boston: Jones and Bartlett), pp.29-47.

Walker, A. and Shipman, P. 1996. The Wisdom of the Bones (New York: Knopf), 338 pp.

Weiner, J. 1994. *The Beak of the Finch* (New York: Knopf), 332 pp.

FURTHER READING

Darwin and Darwinism

Darwin, C. 1859. *The Origin of Species*, 1st Edition (London, John Murray), 502 pp.

Dawkins, R. 1987. *The Blind Watchmaker: Why the Evidence of Evolution Reveals a Universe Without Design* (New York: Norton), 332 pp.

Ghiselin, M.T. 1969. *The Triumph of the Darwinian Method* (Berkeley, CA: University of California Press), 287 pp.

Mayr, E. 1991. *One Long Argument* (Cambridge: Harvard University Press), 195 pp.

Human Evolution

Conroy, G.C. 1997. *Reconstructing Human Origins* (New York: Norton), 557 pp.

Day, M.H. 1986. *Guide to Fossil Man* (Chicago: University of Chicago Press), 432 pp.

Rasmussen, D.T. (Ed.). 1993. *The Origin and Evolution of Humans and Humanness* (Boston: Jones and Bartlett), 146 pp.

Walker, A. and Shipman, P. 1996. *The Wisdom of the Bones* (New York: Knopf), 338 pp.

Evolution

Futuyma, D.J. 1986. *Evolutionary Biology* (Sunderland, PA: Sinauer), 600 pp.

Ridley, M. 1996. *Evolution,* 2nd Edition (Boston: Blackwell), 752 pp.

Skelton, P. 1993. Evolution: *A Biological and Paleontological Approach* (Reading, PA: Addison-Wesley), 1064 pp.

Development and Evolution

Raff, R.A. 1996. *The Shape of Life* (Chicago: University of Chicago Press), 520 pp.

Raff, R.A. and Kaufman, T.C. 1983. *Embryos, Genes and Evolution* (New York: Macmillan), 395 pp.

CHAPTER 4

BEYOND REASON: SCIENCE IN THE MASS MEDIA

JERE H. LIPPS[1]

"We've arranged a global civilization in which most critical elements profoundly depend on science and technology. We have also arranged things so that almost no one understands science and technology. This is a prescription for disaster."

Carl Sagan, 1996

INTRODUCTION

I am outraged by the pseudoscience, antiscience, and plain lack of common sense that I see on television, read in books, magazines, and tabloids, and hear on the radio. All of these distort, confuse, and misinform about science, how science is done, and who practices science. To sell products, much of the media—especially television—exploits the superstitions and fears of unknowledgeable citizens who live in a civilization acutely dependent on science and scientific reasoning. Americans deserve much, much better! That is what I want to show in this chapter.

As a scientist I understand the processes of science and its importance to our well being. As an instructor of some of the best students at one of the premier universities in the United States I see the enormity of the job before us. And as a parent I worry a lot about what this problem means for America's future, just as Carl Sagan did.

One thing is clear to me—we scientists, teachers, and parents have pathetically little help from the **mass media**, television in particular. In fact, they work strongly against us and the public in general, not only against the goal of achieving a **scientifically literate** society but in many other destructive ways already well documented. Why? To make a buck. And the media are afraid to lose even a tiny fraction of their audience by moving from the shock 'em, amaze 'em, fool 'em, and scare 'em approaches to one based on real science that could be just as interesting!

According to Tony Tavares, President of Disney's Anaheim Sports (quoted in *Time Magazine*, August 4, 1997): "Our main goal is to get people to spend their

---

[1]Department of Integrative Biology and Museum of Paleontology, University of California, Berkeley, CA 94720

disposable income with properties associated with the company, whether they're our theme parks, videos, movies, or our sports teams. If you've got a dollar, we want it." How very, very sad. Instead, why didn't he suggest: "Our goal is to earn your dollar with something worthwhile"—why not truth, honesty, and quality, as well as real entertainment, rather than simply more junk to take in more bucks?

For the most part, the people doing this damage to Americans are themselves Americans. Truly this is a prescription for disaster! And their choice of dollars rather than substance, of shock 'em-amaze 'em-fool 'em-scare 'em and make 'em laugh rather than sound science, is not a matter of freedom of speech—it is a matter of responsibility and honor which television and tabloids (as well as some of the other media) have hardly even tried.

In this chapter I briefly lay out the immensity of the problem, cite a few examples, and provide some suggestions on what we can do about the mass media and **science illiteracy**. Much more could be, indeed has been, written about these subjects. Where I differ from some others is that I place significant blame on the mass media. I recognize that the media are controlled by people as ignorant of science as are Americans in general. Hence I accept responsibility, as a scientist, to try to inform people about this problem and suggest solutions.

I see the mass media as the root of the problem because they enthusiastically embrace junk-science, which they know "sells," and are unwilling to risk losing their audiences with real science they don't comprehend. Yet given the opportunity I'd like to work with the mass media because they are not only the problem, they are also the solution—if only they would try! I therefore end this chapter with a proposal to collaborate with them on entertaining or newsworthy programming that properly conveys science so the American, indeed world public gain sound understanding of this crucial aspect of our daily lives.

What Is the Problem?

Ninety-five percent of Americans are scientifically illiterate, according to a concerned Carl Sagan (1996). That means that nearly 200 million people (Table 4.1) in the United States cannot understand how science works, what the process is of evidential reasoning, or how to judge whose opinions can be trusted. Only about 10 million Americans are scientifically literate, most of whom are professional scientists, engineers, or technicians.

TABLE 4.1

Number of Americans (in millions) over the age of 16 who will be scientifically illiterate or literate based on population projections for July 1 of the year indicated (US Census Bureau, 1997) and an estimated present rate of illiteracy of 95% extended to the year 2020.

Year and Total US Population	Scientifically Illiterate (95%) over age 16	Scientifically Literate (5%) over age 16
1998: 270+	197+	10+
2000: 274.6	201+	10.5+
2010: 297.7	222+	11.6+
2020: 322.7	240+	12.6+

Because the degree of science illiteracy has been estimated to lie between 93 and 97%, depending on how it has been measured, these numbers are approximate.

TABLE 4.2

What science facts do Americans know?

Statement or question	Percentage of Americans agreeing with the statement or who could provide the correct answer
The center of the earth is very hot.	78%
Radioactivity is both man–made and natural	28%
The oxygen we breath comes from plants	85%
The earth goes around the sun once a year	47%
The universe began with a huge explosion	35%
Humans developed from an earlier species of animal	44%
Cigarette smoking causes cancer	91%
Humans did not live with dinosaurs	52%
Antibiotics kill bacteria but not viruses	60%
What is DNA?	21% gave a reasonable answer
What is a molecule?	9% gave a reasonable answer

Results of the National Science Board's (1996) survey of American's grasp of science. The survey (2,006 adults, selected randomly) included 10 questions about science facts. Three-quarters of Americans could not pass the quiz. The first eight questions were multiple choice and the last two were simple word answers. The correct statements listed here have been reworded.

While the degree of science literacy is not easily determined, the National Science Board's (NSB) 1996 survey of the American population indicates that only 2% understand how scientific theories are developed and tested, only 23% are minimally able to explain the nature of scientific study. Americans are not much better when it comes to scientific facts. Some 75% fail a rather simple test of eight multiple-choice and two long answer questions (Table 4.2; National Science Board, 1996a). This just is not good enough (Ehrlich and Ehrlich, 1996; Lederman, 1996; Sagan, 1996).

Science illiteracy contributes to an antiscience mentality that threatens our very existence (Ehrlich and Ehrlich, 1996). This "prescription for disaster" may well have everlasting consequences for our country. Science is ubiquitous in America and many decisions, both national and personal, depend on understanding of it. We, as a nation, cannot tolerate widespread science illiteracy.

What are we doing to correct the situation? The NSB (1996b) recommended that "The Nation must put absolute priority on educating and training all members of society in mathematics, science, and engineering so they may be productively employed in an increasingly sophisticated global economy. This educational process is a lifelong endeavor . . . ."

Indeed, many are working hard to improve science education—the President and Vice President of the United States, the National Science Foundation, the National Academy of Sciences, the National Science Teachers Association, and a multitude of other people, organizations, scientific societies, and institutions. As a result of these efforts we are making some progress in our schools. And these efforts pay off, for the NSB report shows that education and science literacy go hand-in-hand—the more education a person has, the better that person understands science. However heartening this progress in education may seem, it is offset, diluted, even erased in the general public by the most ubiquitous element of American life, the mass media, television in particular. Even in colleges and universities, for example, knowledge and reason are being degraded by the pervasive influence on college-age students of entertainment and sound bites (Sacks, 1996). But life is not "entertainment" and clear reasoning requires more than simple "sound bites," so we have a general public woefully unprepared for aspects of everyday living.

Science illiteracy is not confined to America, of course—it is a global problem. While some countries may do a somewhat better job of science education than the US, most accomplish much less. Data are not available for science illiteracy worldwide but if the American figure of about 95% is taken as a minimum, then there are billions of people across the globe (Table 4.3) who are unable to understand a large portion of what effects them.

If science is important to America, it is just as important, if not more so, to countries which face a multitude of scientific problems in even larger measure than the United States—population growth, environmental deterioration, biodiversity decline, greenhouse effects, health problems, food and air quality, natural hazards, and many others. And the consequences for most of the rest of the world are significantly greater than for America. Pesticides, toxic chemicals, tobacco, and false health remedies are foisted on the populaces of many other countries in greater amounts than in the U.S. because of less stringent rules and regulations and limited public understanding of their effects. Were their populations more knowledgeable, countries around the world could more effectively deal with these problems. Many of the problems are not constrained by political boundaries and so threaten even countries with substantial science literacy, if not the entire world.

Are the approximately 10 million scientifically literate people in America enough to safeguard the other 270 million inhabitants of the United States? Are the probably many fewer than 200 million worldwide enough to assist the other nearly 5,800 million global inhabitants deal with scientific problems? Probably not, for people are faced with individual scientific decisions almost daily and many must vote or otherwise decide about scientific matters affecting their region or country, choose political representatives who should also be able to understand scientific problems, and seek satisfying employment that is increasingly dependent on basic

TABLE 4.3

Number of people (in millions) over the age of 15 worldwide who will be scientifically illiterate or literate based on population projections for mid-years as shown (McDevitt, 1996) and assuming a degree of science illiteracy identical to that estimated for the United States.

Year and Total World Population	Scientifically Illiterate (95%) over age 15	Scientifically Literate (5%) over age 15
1998: 5771	3700+	199+
2000: 6090	4000+	213+
2010: 6861	4700+	250+
2020: 7599	5400+	285+

This percentage (95%) is undoubtedly too low and can be expected to increase through time because the increasing populations of most countries will tax their educational systems even more than they do now.

TABLE 4.4

Although there are many sound reasons for rejecting pseudoscience, its popularity in America appears to be increasing as the new millennium approaches.

Pseudoscience	Reason for Rejection
Astrology	Ancient pagan beliefs
Crystal worship	No evidence; beliefs
Creationism	Beliefs
Numerology	Failed tests
Phrenology	Insufficient evidence; failed tests
Psychic Healing	Insufficient evidence
Psychic Prediction	Disproved by failure to perform
UFOlogy	Insufficient evidence

scientific knowledge. *The answer then surely is that most people everywhere should be able to understand the basics of the scientific process.*

WHAT DO WE KNOW ABOUT SCIENCE?

What Is Science?

A fundamental misunderstanding virtually everywhere in the world is that material in the mass media presented in a scientific manner is real science. Unfortunately, it seldom is. Instead it is largely pseudoscience, antiscience, superstition, and dogma (Table 4.4). This deluge contributes hugely to science illiteracy by confusing fact with fiction, scientific theory with belief, and scientists with nonscientists. Why people are fascinated with and willingly pay money for pseudoscientific or antiscientific claims and products is a deep problem, but among the contributing factors are poor education, personal and mass delusion, indoctrination, hopelessness, fear of other people, apprehension about the world around them, and dread of the unfamiliar (Eve and Harrold, 1991; Miller, 1987; Shermer, 1997). The mass media too commonly capitalize on these to increase their audiences. The answer lies in education by all means possible—schools, person-to-person, political bodies, corporations, and especially the mass media.

Antiscience, pseudoscience, and weird beliefs are all around us—on TV and radio; in movies, books, newspapers, magazines; from street corners to pulpits to meetings of school boards and other political bodies; even in colleges and universities (Ehrlich and Ehrlich, 1996; Sacks, 1996; Shermer, 1997). Exposure to real science pales in comparison (with NASA's 1997 Pathfinder landing on Mars a welcome exception). **Antiscience** simply ignores scientific reasoning in making its claims. **Pseudoscience** makes claims that have the trappings of science and sound scientific but disallows proper tests of its claims which are based on inadequate evidence, false authority, or unsupported beliefs. The results can be humorous and entertaining, but sometimes are tragic and costly. For example, Americans spend billions of dollars each year on peudoscientific remedies for health problems.

Science, on the other hand, employs logic, critical thinking, and appropriate evidence; subjects all authority to scrutiny; and encourages testing of its claims. These are the basics of science, and almost anyone can learn them.

While some tout the "**scientific method**" as the way science is done, the formalized step-by-step process taught in many classrooms is a reordered version of the actual activities most scientists go through. This formalized version of the scientific thinking process appears dull and agonizing and is one reason why some people dislike or even fear science.

But science actually can be more like solving a puzzle or playing a game that is enormously good fun. Few scientists work by the formalized "scientific method." Instead, they become excited, hopeful, interested, intrigued, amazed, puzzled. Their ideas come to them in the shower, on the freeway, while playing baseball with their kids, as well as in the laboratory or library. The best science is creative and exciting. It certainly can be just as fun and entertaining as pseudoscience. In most cases, it is more so. And it provides a challenge and reward to get it right. A formalized blueprint to the methods of science is not necessary—to understand how science is done or practice it on a personal basis merely requires a critical and evidential approach to thinking, as will be detailed later.

The extraterrestrial Unidentified Flying Object (UFO) phenomenon is a good example of a clash between science and pseudoscience. A large segment of the scientific community readily accepts a high probability for the existence of extraterrestrial (including intelligent) life somewhere in the Universe, but few scientists accept UFOs. The photographs of purported UFOs; unsupported personal accounts of sightings, visitations, abductions, sex with and sexual experimentation by aliens; and evidence more logically explained by known earthly phenomena are woefully inadequate to support such an important and momentous claim as alien spaceships visiting Earth. This is a serious claim that—if true—would have enormous impact on our lives and civilization. But no one has ever produced any evidence to support it unequivocally—no space ships, aliens, or parts or pieces of either for legitimate scientists to examine.

UFO claimants, of course, have explanations to account for science's reluctance to accept their claims: The government is covering up evidence (such as crash-landed spaceships and bodies of aliens); the scientific community is conspiring to suppress the facts; scientists are biased and untrustworthy; or the extraterrestrials have modified the minds of people who encountered (or were abducted by) them, altering or wiping out their memories (Corso, 1997; Fawcett and Greenwood, 1992; Marrs, 1997). The claimants are always able to come up with one reason or another why scientists and other disbelievers are not to be trusted, no matter what the pseudoscientific or antiscientific claim may be. This distrust of the scientific community is so commonly offered as an explanation for the nonacceptance of claimants' views that it is a sure index that the claim is suspect.

Belief in UFOs, as with other pseudoscientific claims, runs deep (Shermer, 1997). UFOs are touted at innumerable sites on the World Wide Web and in a multitude of books, magazines, and newsletters. Tens of thousands of people attend celebrations about them where a good deal of money is spent on UFO books, trinkets, shirts, and so forth. Take a look at the selection of volumes on UFOs in your local bookstore—there are literally hundreds which not only promote claims that cannot be supported by firm evidence but assert that the "truth" about UFOs has somehow been "covered-up."

Once in a while such claims lead to tragedy. Recently, for example, 39 members of the Heaven's Gate cult believed they would be miraculously transported to a spaceship following Comet Hale-Bopp if they all committed suicide. They did just that in the spring of 1997, although there was no evidence of anything following the comet, other than its tail. It is sad that so many people could be so deceived, but their beliefs were reinforced by far too many television programs, articles in tabloids, and books all trumpeting the same pseudoscientific views of UFO visits and crashes, alien autopsies, abductions by extraterrestrials, and sexual experimentation on humans by aliens.

Credence was lent these UFO myths by a wholly unexpected source—the insurance industry. The London insurance firm of Goodfellow, Rebecca, Ingrams, and Pearson offered policies covering abduction, impregnation, and death caused by aliens (Underwriter's Report, April 10, 1997). Coverage was bought by several hundred people, among them the Heaven's Gate cult which paid $1,000 for coverage of 50 of its members. In the aftermath of the mass suicide at least one California lawyer thinks the cult has a claim—after all, he notes, no one can prove the members were not abducted. The surviving cult members would be entitled to $39,000,000. A spokesperson for the London firm noted that the company no longer issues alien-related coverage because "We don't want to contribute to a repetition of the Heaven's Gate deaths" (Underwriter's Report, April 10, 1997).

Even the State of Nevada has jumped into the UFO act. Largely in response to repeatedly reported sightings of UFOs along a desolate 98-mile stretch of Route 375, Nevada officially named it the "Extraterrestrial Highway" and erected signs depicting aliens and alien spacecraft (Wheeler, 1996). The highway passes near a classified US Airbase which the UFO'ers claim conveniently covers up the UFO activity. They even maintain that alien bodies are stored at the base. The Extraterrestrial Highway is now heavily promoted by the Nevada Commission on Tourism with advertisements like the one that appeared in *Sunset Magazine* in September 1996. [Ironically, the ET Highway actually passes through Devonian-age rocks that on the basis of good scientific evidence are thought to have been formed by an extraterrestrial impact about 360 million years ago (Warme and Sandberg, 1996; Leroux, Warme and Doukhan, 1995). There may be some truth to the highway's name, after all!]

To some, science can be frightening and remote. It commonly deals with esoteric and complicated subjects, and some of its most obvious technical manifestations (nuclear power, for example) have caused significant problems. Even in public forums, many scientists use technical jargon that alienates, graphs and charts that are all but uncomprehensible, and illustrations unrecognizable to a layperson. Science's best known practitioner, Albert Einstein, is regarded as a genius who had knowledge unattainable by an average person. And we scientists tend to criticize, and sometimes penalize, those among us who do try to make science understandable to the public. No wonder that the mass media portrays scientists as weirdoes, nerds, crazies, even evil-doers (Figure 4.1).

Yet in spite of the public image of scientists and widespread science illiteracy, 40% of Americans have a "high interest" in scientific discoveries and 70% in new medical findings (National Science Board, 1996). Interest in things scientific is large and provides potential for increasing science literacy, if proper means are devised.

Science Literacy

Science literacy can be defined in several ways. One way involves three simple elements, the abilities to (1) think critically, (2) use evidential reasoning to draw conclusions, and (3) evaluate scientific authority. Science literacy does not necessarily mean the accumulation of scientific facts, although a foundation of basic factual knowledge seems required.

Critical thinking can be learned by practically anyone. Important in scientific reasoning it is even more so in daily living, for a clear understanding of our

FIGURE 4.1   Scientists commonly are regarded as nerds, weirdoes, or even evil-doers by the general public. Actually they are ordinary people with ordinary concerns, emotions, and fears who have learned to use critical thinking and evidential reasoning to understand their surroundings. They are usually enamored with their jobs and in this way may differ from some others. (A cartoon by Jeremy C. Lipps.)

surroundings and problems makes life enjoyable and safer. All of us make assumptions, hold biases based on previous experience, and have emotions. Because these interfere with our ability to interpret the world around us they must be set aside if we are to properly evaluate information and solve problems. As a matter of self-defense against those who would victimize us with hoaxes, frauds, flaky schemes, or do us physical harm, critical thinking is essential.

Critical thinking involves eight skills (Table 4.5) that help evaluate situations rationally. These promote clear and creative thinking that commonly simplifies the solutions both to daily and larger problems. Perhaps the most important skill is to begin by asking questions. These should be sound and relevant, questions that can be tested by evidence rather than explained away by assumptions, biases, or emotions. Consider other possibilities and their testable answers, and accept the fact that not all questions have immediately obvious clear-cut answers and that as knowledge improves over time, understanding will change. Uncertainty and change are assured!

Evidential reasoning—six rules (Table 4.6) by which to examine the veracity of any claim—can also be learned. I consider these rules not only the basis of the

TABLE 4.5

Skills for critical thinking (from Wade and Travis, 1990).

Skills	Simple Techniques
1. Ask questions: be willing to wonder	Start by asking "Why?"
2. Define the problem	Restate the issue several different ways so it is clear.
3. Examine the evidence	Ask what evidence supports or refutes the claim. Is it reliable?
4. Analyze assumptions and biases	List the evidence on which each part of the argument based. The assumptions and biases will be unsupported..
5. Avoid emotional reasoning	Identify emotional influence and "gut feelings" in the arguments, and exclude them.
6. Don't oversimplify	Do not allow generalization from too little evidence.
7. Consider other interpretations	Make sure alternate views are included in the discussion.
8. Tolerate uncertainty	Be ready to accept tentative answers when evidence is incomplete, and new answers when further evidence warrants them.

These can be acquired readily by most people.

way scientists work but as a guide to intelligent living. The first steps are to ask whether the evidence presented is sufficient to support the claim, and whether other evidence (including that which might be gathered using different avenues of investigation) does or could disprove it. Both of these steps require skill in critical reasoning. And because personal biases must be set aside you must ask yourself whether you are being honest and objective in your evaluation of the evidence rather than deceiving yourself.

TABLE 4.6

Rules for evidential reasoning (from Lett, 1990).

Falsifiability	Conceive of all evidence that would prove the claim false
Logic	Argument must be sound
Comprehensiveness	Must use all the available evidence
Honesty	Evaluate evidence without self-deception
Replicability	Evidence must be repeatable
Sufficiency	1. Burden of proof rests on the claimant. 2. Extraordinary claims require extraordinary evidence. 3. Authority and/or testimony is always inadequate

All claims should be subjected to these rules which serve as a guide to intelligent living and the scientific method.

The evidential underpinning of a claim also must be carefully evaluated. Have the claimed observations or experiments been repeated and have they yielded the same result each time? Are the supporting lines of evidence logical, a series of natural consequences leading from one to another and yet another in an understandably rational manner? Logic is often presented in formal and formidable ways, and although it requires critical and clear thinking it really is not so difficult to fathom (Perkins, 1995).

The logic of a claim is crucially important and you should be especially wary of claims that seem illogical or defy common sense. To be accepted, such extraordinary claims require the backing of very strong abundant evidence. For example, fuzzy photographs of shiny discs in the sky or reports of scorched ground or marks on soil would be insufficient to support claims of extraterrestrial visitors. Any such claim would probably require no less than evidence of the aliens themselves—a live specimen, a body, an appendage, or a DNA test of tissue samples—and the evidence would have to be made available to the scientific community for repeated testing so that spurious results could ruled out. Support less than this would be regarded insufficient because it could have numerous alternative explanations (natural or man-made phenomena, mistaken identity, delusions of the observer, hoaxes, and so forth).

Lastly, the claimant's reasoning and expertise must be scrutinized. Claims should not be accepted solely on the basis of testimonials from so-called experts. Anyone can say anything! The crucial factor is the evidence, the facts supporting the claim, not the person making the claim.

Like the evaluation of evidence itself, that of scientific authority depends on critical thinking and evidential reasoning so apply these to evaluate the claimant as well as the evidence presented. Be skeptical. Does the so-called authority use critical thinking and evidential reasoning? Are the claims presented without undue call on unsupported or unsupportable arguments?

One generally quite effective way to evaluate the validity of claims made by so-called authorities is to determine whether they have passed the test of **peer-review** (a scrutiny for completeness and accuracy by members of the scientific community) and been published in recognized scientific journals. Despite the fact that many "authorities" of dubious credentials make the claim that the scientific community has "suppressed" their work, prohibiting its publication, their results are in fact very seldom even submitted for publication in the customary peer-reviewed scientific journals (Scott and Cole, 1985).

With regard to expertise, does the claimant have the appropriate experience and background to make the claim? Most true authorities in science have positions that allow them to practice science actively, usually in an institute, corporation, university, or governmental agency engaged in scientific research. False authorities often lack such affiliation and claim to be authorities by virtue of bogus degrees, degrees from institutions of questionable academic standing, authorship of purportedly "authoritative" books or pamphlets, or affiliation with nonscientific or pseudoscientific organizations that nevertheless have scientific-sounding names. Some simply proclaim themselves authorities on the basis of personal experience and self-training (a path to knowledge once common but now quite rare).

SCIENCE IN THE MASS MEDIA

The mass media have a good deal of difficulty dealing with science, yet their influence is enormous. They assail us daily with information, some reliable, some much less so. How do they do with science?

Unfortunately, many of the people responsible for the information content of the mass media are no better informed about science and its processes than is the general public. Most newspaper editors, for example, seem to understand science about as well (or poorly) as the public at large—about the same percentage of both groups believe that humans and dinosaurs coexisted in the geologic past and uncritically accept other misconceptions as well. How can editors of newspapers prepare stories for publication on such subjects when they are totally unable to distinguish good science from bad, truth from fiction? Screen writers are not any better, for although most would like to present intelligent stories about science they simply lack the basic knowledge to do so (Steve Allen, personal communication, 1997). Because the people in charge, as well as those reporting news or crafting stories have little scientific understanding, the mass media serve science very poorly!

The influence of the mass media varies from medium to medium and the public's use of each. I know of no in-depth evaluation and comparison of the way science is presented in the various media so I give my own impression in Table 4.7. There are many potential pitfalls in making such an assessment, perhaps foremost the mistake of assuming that what I as an evaluator of a medium read, view, or listen to is the same as the general public encounters. Many scientists, for example, have told me they regard television coverage of science as generally "good"—but they selectively watch only science-related programs that for the most part are viewed by a very small percentage (less than 5%) of the viewing public and that by "preaching to the choir" have little impact. I have therefore attempted in Table 4.7 to give a broad-based assessment of the situation, but a proper evaluation must be done by general surveys of American viewers, not by what individuals think. When this is done, the problem facing America will be seen to be obviously very great.

Here I separate the mass media into two categories: Active and Passive. The **Active Mass Media** require considerable effort on the part of the public to absorb the message presented. I include in this category books, newspapers, magazines, tabloids, and the Internet, use of which require a person to make a positive decision to engage the medium and to then actively participate throughout the encounter. I contrast these with the **Passive Mass Media**—television, radio, and movies, for example—which require a minimum of decision-making on the part of the recipient for whom the reception of the ideas presented is entirely passive.

Gradations, of course, exist between these categories and among their components. For example, although many people choose books or magazines quite carefully and comprehend them well, others use them in a more passive way, select-

TABLE 4.7

A personal impression of the relative influence (denoted by the number of "+" symbols) **on the general public and the quality of science coverage in various media** (where the number of "+" symbols indicates the relative content of high-quality coverage; that of "-" symbols, low-quality coverage).

Medium	Influence	Science
Television	++++++	- - - - - +
Newspapers	++++	- - + +
Tabloids	+++	- - - - -
Movies	++	- - - - +
Magazines	++	- +
Internet	++	- - - + +
Radio	++	- - - +
Books	+	- +

ing them without much thought and flipping through their pages; and although many use television as a passive medium, others are very choosy about the programming they view and may think deeply about what they have seen.

Active Mass Media

Americans read a great deal each year, and of course select their reading matter—books, magazines, newspapers, tabloids, and Internet sites, or articles in them—based on personal interests. Those who like romance novels or stories about paranormal phenomena are unlikely to read books or articles about science unless they find it appealing also. In all bookstores I have visited (well over 100 in several cities) I have found that books and magazines devoted to pseudoscience and the paranormal vastly outnumber those dealing with real science, evidently because these topics attract much larger audiences. Indeed, some bookstores are dedicated solely to the paranormal. Why are pseudoscience and the paranormal so popular? Probably because these kinds of topics are easy for untrained people to relate to whereas real science is not. But the situation need not be this way. Science can be presented in a compelling fashion—in dramatic, funny, and engaging ways dealing with topics of broad interest. But to do so requires knowledge of science and its processes!

Science coverage in popular magazines is highly variable Some of the most popular seldom present science at all while stories or columns on the paranormal appear regularly. *Time, U. S. News and World Report,* and *Newsweek* often include science stories although they also regularly report on pseudoscientific topics without enough critical comment. All in all, millions of people read popular magazines, but they are far less influential than other media which reach even more people more easily.

Newspapers, particularly those in large cities, seem to do a fairly effective job of science reporting, especially when the writers are trained in science or are dedicated to science writing only. But general reporters often seem even less knowledgeable about science than most of the public—they work under the pressure of deadlines and must complete their write-ups even if they lack basic understanding of the science they are assigned to report. Too commonly they ask irrelevant questions; probe for "newsworthy" yet nonexistent controversies about the meaning of scientific data; plumb for conflicts that do not exist between researchers or research teams; and distort, sometimes badly, the actual science news.

Tabloids for sale at grocery store checkout stands clearly aim to do one thing—make a lot of money for their publishers. They sensationalize everything, including what little science appears in their pages. Science may be denigrated or lauded, depending on the slant the publishers think will sell the most copies. While many view these papers as light entertainment, some of the tabloids have become rather successful in investigative reporting of particularly sensational events (such as the O. J. Simpson trial). Such successes increase enormously the apparent credibility of other stories in the tabloid, even those that obviously are beyond belief.

Use of the Internet via the World Wide Web to access information is highly selective. Millions of sites exist on the WWW but few are accessed by large numbers of people. Our own Berkeley Museum of Paleontology site (http://www.ucmp. berkeley.edu), one of the first on the Web and containing over 3,000 pages of material on paleontology, is accessed about two million times a month. The actual number of users is much less (because many access the site repeatedly by following links to other pages), and most other sites devoted to science are used by even fewer people. In any case, the number accessing any of these sites is paltry compared to some nonscience sites such as, for example, the Playboy site which according to the Playboy Webmaster (personal communication, 1996) is accessed some 15 million times each day!

The more than 50 million people who use the Web can be influenced only by sites they choose to access, largely those that match their particular interests. And since much use of the Internet depends on searches made by key words, there may be even more selectivity in topics sampled than is typical for readers of newspapers or magazines. WWW sites focused on the paranormal vastly outnumber those devoted to real science so I think it unlikely the Web will contribute much to increasing public science literacy—because the Web is so cluttered with sites, individuals will continue to choose those of special interest to them and the paranormal is far more abundantly represented than science.

The picture is not entirely bleak. Martian Pathfinder WWW sites exceeded 550 million hits in the first few weeks after the landing on Mars on July 4, 1997, an unusually large number. Among magazines, *National Geographic*, with a circulation of more than 9 million, includes many scientific articles; *Discover Magazine* (circulation 1.2 million) deals with science and debunks pseudoscience; and even the major news magazines occasionally have excellent science articles. Carl Sagan's book *Cosmos* was bought widely, chiefly because it intermeshed with his exceedingly popular television series, and his other books have sold well too, based on the public's awareness of him as a television personality and award-winning author.

Yet authors who promote antiscientific or pseudoscientific views have readerships as large or even larger than Sagan's, and a huge number of books, magazines, and articles are devoted to these views. Rush Limbaugh, for instance, sold over 7 million copies of his books (1992, 1993) which remained on the *New York Times* "best seller list" for many weeks in spite of an abundance of antiscientific and pseudoscientific misstatements, errors, and logical inconsistencies (Perkins, 1995).

In general, the active media—books, newspapers, magazines, tabloids, and the Internet—reinforce previously entrenched interests and knowledge and do not attract new readers. And compared to the passive media the number of users is very low. Whether good, bad, or indifferent to science, the active media can have only relatively little influence in changing public awareness of science and improving science literacy.

Passive Mass Media

Radio, movies, and television provide only the most limited understanding of science. Today, radio is mostly music and talk shows, few of which either practice or preach critical thinking and evidential reasoning. Some, such as the *Rush Limbaugh Show*, actually disparage science and scientists almost daily. Limbaugh's listeners number over 20 million. This is significant. Other talk show hosts—such as Dr. Dean Edell who often and quite effectively debunks pseudoscience and antiscience—have large audiences too, but it is unlikely they undo the harm of pseudoscience zealots.

Movies deal with many science-centered topics, from dinosaurs to aliens, tornadoes to earthquakes. Because most people go to movies specifically to be entertained, not to learn, movies on such topics are probably not too detrimental to science understanding. Yet movies have influence. They give an impression of who scientists are (nerds, flakes, sometimes bad guys); what they do (crazy stuff, usually); and why they do it (to conquer the world and win the girl). But they almost never show the processes of science. And some reinforce pseudoscientific themes. I took my son to see *Independence Day,* the fantasy film about a massive alien invasion where whole cities were destroyed (including ours—my son and I *knew* this was a fantasy!). Yet when the alleged Roswell UFO crash and capture of extraterrestrials came on the screen a woman behind me excitedly told her three children, "That's true! I saw it on television!!"

To me, television is truly the worst of the media because it reaches so many homes and is so easily absorbed passively that it is very influential, often to the detriment of science (Figure 4.2).

TV has little to be proud of when it comes to science. Most children and many adults uncritically watch many hours of television weekly. The average kid in America spends more time watching TV than he or she does in school! It is therefore particularly unfortunate that science is so poorly presented. People learn from television, their views and behavior are commonly shaped by what they see. To protect the industry from lawsuits, television executives claim their programs do not influence people. Yet when these same executives seek to sell billions of dollars of advertising they make just the opposite claim. It can't be both ways.

FIGURE 4.2 Television is mesmerizing, passive, and highly influential. (A cartoon by Jeremy C. Lipps.)

Television distorts and misinforms by mixing programming. Entertainment programs look like news programs while news programs look like entertainment. Documentaries are designed to be entertaining, not necessarily factual. Viewers cannot easily distinguish reality from fantasy.

Moreover, the networks seem to uncritically accept and buy programs that distort, lie, and cheat about science, and promote antiscientific or pseudoscientific beliefs. Anyone with a slick program on any outrageous subject can seemingly find a network that will buy and show it, a situation that gets worse as time goes on. To most scientists and educated laypersons these programs are ludicrous, farcical, not worth the time to view them. "Trash," we'd say. Yet even in reruns these programs can reach as many as 7 million viewers, and millions more than that in their first showings (Emery, 1997).

NBC, for example, twice aired *The Mysterious Origins of Man*, a program full of factual errors and distorted and spurious theories and featuring self-proclaimed authorities claiming "scientific cover-ups." The program outraged knowledgeable scientists. Young people were confused and their teachers were confronted with erroneous questions.

Some of the more obvious errors aired in the program were assertions that human remains 55 million years old were found at Table Mountain in California and that humans walked with dinosaurs, said to be shown by trackways preserved in Cretaceous (more than 65 million year-old) mudstones in Texas. The program claimed that scientists covered up these "facts" in a gigantic conspiracy. Far from it! Science long ago showed that the supposed great age of the California remains is incorrect, that they in fact come from a near-modern cavelike shelter (Blake, 1899), and that the "human" footprints preserved alongside those of dinosaurs are actually weathered dinosaur prints or carved fakes (Hastings, 1987). The program's writers and producers, themselves nonscientists, either did poor job researching the facts or ignored the established findings intentionally. Yet they dredged up these falsehoods, packaged them in an entertaining fashion, and had them presented to the public by an authoritative-sounding Charlton Heston.

Other demonstrably erroneous theories were presented, such as the incredible idea that during the most recent ice ages the crust of the Earth abruptly slipped some 2,000 miles, carrying mammoths from temperate regions to the Arctic so fast they didn't have time to swallow or spit out buttercups they were eating. Alternate explanations, such as the fact that buttercups were prevalent in the Arctic when and where the mammoths died, were never mentioned. The basic process of science was completely distorted.

The "scientific authorities" shown on the program as backing such claims lack proper credentials, are not employed as scientists, and neither publish in the peer-reviewed scientific literature nor practice the scientific method. One so-called expert on the Cretaceous "human" footprints and referred to in the program as "Doctor" received his doctorate degree in anthropology from an unaccredited "college" with a "campus" consisting of a single house next to the Sherwood Baptist Church in Irving, Texas. This "educational institution" has neither laboratories nor faculty and its president and conferrer of degrees is none other than the "Doctor" himself (Kuban, 1989). An expert authority? Probably not!

The Mysterious Origins of Man should have been labeled fiction or entertainment instead of science, but like other networks NBC seems unable or unwilling to evaluate properly the scientific validity of its programs. Sadly, numerous other programs, some regular features, continue to convey the same sorts of distorted and false messages about science.

Sometimes television devotes whole series to such pseudoscience programming. For instance, the entire week of March 24, 1997, was declared "Alien Invasion Week" on the Learning Channel—an odd juxtaposition of terms! Each night featured pseudoscientific programs on UFOs, alien abductions, alien autopsies, and similar topics. While most of the programs included a sprinkling of mild disclaimers such as "might have been," "could have been," "alleged," these were surrounded by images of flying saucers, big-eyed aliens, and authoritative-sounding "experts," all accompanied by dramatic music. In a few cases a skeptic or scientist waxed forth on why the notions presented fell short of acceptable scientific standards, but during these segments the music and images changed to a decidedly less dramatic and less engaging tenor.

Although UFOs may well exist, lights in the sky, Biblical tales, fuzzy recollections of years long past, blurry photographs, and outright fakes are hardly the kinds of evidence that such claims require. Aired on the Learning Channel, this programming does not encourage learning in any sense of the word but rather indoctrination. Yet programs of this type have enormous influence—they reach millions of people who in the absence of information to the contrary are encouraged to believe that "if it's shown on television, it must be true!"

It seems to me likely that most American viewers have forgotten the lessons they once learned in school about how to reason and think scientifically. Faced continuously with the trash-science of TV, they tend to accept it uncritically, unthinkingly, and without recalling what is and what is not acceptable in even normal logical reasoning. And by eroding the valuable lessons once learned, TV's version of "science" actually promotes science illiteracy. This is more than just a prescription for disaster—it is a great waste of time, effort, and money, and provides a center stage for charlatans to foist half-truths, mistruths, and outright lies on the American public. TV's trash-science is truly a major contributor to "the dumbing of America."

But the problem does not stop at America's shores. The United States feeds TV programming to the rest of the world. In Australia, for example, many programs are shown soon after they appear on American screens. And so it is, worldwide. The problem is not just the dumbing of America, it is "the dumbing of the World!" Across the globe, societies are struggling to ensure that their citizens are increasingly literate and scientifically useful. In fact, many of America's megalithic companies now move offshore not only to decrease costs but to attract a workforce appropriately educated in science and technology. The State of California already faces the loss of high-technology companies which cannot find enough workers sufficiently educated about even simple scientific and technological concepts (*Oakland Tribune*, 1996).

While TV may be a "vast wasteland," it and the other passive media do present good science programs once in awhile. *National Geographic Specials, Nova,* and similar TV programming are excellent examples. And the format followed in the *Jacques Cousteau Specials* generally does a fine job of showing how science is done—Cousteau identifies a problem, one or a few hypotheses are proposed, and his divers then go out to gather scientific data to solve the problem. Although the programs tend to be a bit overly dramatic and at times take poetic license, they have been effective teachers of science around the world.

But other programs, especially those shown on educational channels such as PBS and regarded widely as presenting science fairly well, fail to provide an antidote to the trash-science offered elsewhere on TV. Most of these are merely films of scientific subjects, commonly various animals or plants, with a monotonous voice-over describing the scene. Rarely is the *process* of science part of the program. The thrill of discovery, the sorting of hypotheses, the use of logic, evidential reasoning, and critical thinking—almost never are these truly educational aspects of science included.

The shoddy treatment of science on American television today cannot be accepted. At the very least we should expect clear thinking, sound reasoning, and real science to be as familiar on television as the sex, violence, and sensationalized pseudoscience that now fills the airways. If Americans can learn about antiscience, pseudoscience, sex, and violence from TV, they also can learn about real science if only television would take the responsibility and provide the needed programming.

How might this be done? Understanding science is not especially difficult; the process and activities of science and scientists are often dramatic and exciting, sometimes hilarious; and since scientists are real people with the same strengths and weaknesses as everyone else they get themselves into the same situations that most people do. Drama, excitement, humor, humanness—these of course are the fodder of TV programs as we already know them. Science is perfect for prime time! In fact, Leon Lederman has already charted the way toward production of a dramatic prime time TV program featuring science and scientists (Hirshon and Lederman, 1996). This is a good first step that I hope will lead to incorporation of science and its reasoning and methods into numerous regular prime time programs, including situation comedies.

WHAT CAN WE DO ABOUT SCIENCE ILLITERACY?

Those of us who are scientists, educators, and parents can do much to improve the situation. Our chief target should be proper representation of the *process of science* in the mass media, especially television. We need to think creatively how best to achieve this goal, and must certainly become more proactive. Our suggestions and criticisms need to be presented constructively and in a way that does not threaten the monetary bottom line of mass media corporations. It would be to our advantage, for example, to convince the corporation decision-makers that real science "sells"; that it is human—dramatic, exciting, funny, poignant, tragic; and that it therefore is choice material for well-received programming in prime time. There is, in fact, excellent material in science that would fit prime time programming hand-in-glove, and good reason to believe that significant progress can be made toward correcting illiteracy in science, if we choose to accept the challenge. If the mass media can so effectively peddle trash-science, they can promote good science just as easily (Figure 4.3).

Here I list a few of my ideas for how scientists can become more proactive, but anyone concerned with science literacy and the future of societies can contribute to the solution. You only need to think critically, evidentially, and creatively to come up with other ideas based on your own background and situation. Then put your ideas into action!

Here are mine for my fellow scientists:

1. *Join the battle, in spirit if not in action.* Scientists commonly disparage colleagues who interact actively with the public and sometimes even penalize them by inhibiting their promotions, election to prestigious organizations, and awards of grants, honors, and the like. Since most science in America and elsewhere is supported by taxpayers, this behavior seems both inappropriate and ultimately self-defeating. Instead of derogatory statements and actions, scientists should heap praise and support on colleagues who attempt to meet science's obligation to the

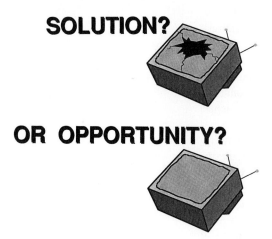

SOLUTION?

OR OPPORTUNITY?

Figure 4.3   The solution is not to destroy or ignore the mass media, but to use them to enhance science literacy.

public. Not every scientist needs to be active, but some certainly must be.

2. *Give stimulating and exciting public lectures,* especially to children's groups. Many scientists are more than pleased to do this, and it helps—exposure to good science at a young age can lay a solid foundation for future growth.

3. *Interact with journalists.* An easy way to do this is to rely on the public relations officers that many universities and scientific institutions already employ. News organizations will almost always work on and publish stories coming from such sources.

4. *Work with television writers and producers* to get good exciting science on television and in the movies. Many of these people have little or no contact with scientists and real science and so embrace pseudoscience for lack of knowledge. Scientists need to work with the television industry, not simply criticize from the outside (as I do in this chapter!).

5. *Write about real science for TV, movies, magazines, and newspapers.* Television may be a difficult market to break into, but many newspapers will gladly publish a column or story from a scientist, especially on a timely topic of general interest.

6. *Inform your local TV, movie, book, or radio reviewer when you see bad science.* Let them know what is real science and what is not in programs, films, books, and so forth. The reviewers are always on the lookout for a good story idea and you will be helping them to do a better job informing the public.

7. *Ask the TV Academy of Arts and Sciences to institute a "best science" category in the Emmy Awards.* There are already many categories—lighting, numerous different kinds of music, and so forth. Why not just one to encourage excellence in TV science?

8. *Urge the networks and your local TV stations to consult or employ scientists to assist the evaluation of programs making scientific claims.* If programs are not real science, label them fiction!

9. *Issue statements about bad science and pseudoscience to the media.* The Committee for the Scientific Investigation of Claims of the Paranormal (CSICP) has formed a Council for Media Integrity that holds press conferences and issues awards. It gave its 1997 "Candle in the Dark" Award (after Carl Sagan's 1996 book subtitle) for excellence in television science to *Bill Nye the Science Guy,* a program

aimed at youngsters. It awarded its 1997 "Snuffed Candle in the Dark"—for the worst portrayal of science—to Dan Aykroyd for his production of the seemingly fact-based but actually super-pseudoscientific program *Psi Factor* (see: http://www.psifactor.com/).

10. *Invest money.* The CSICP Council for Media Integrity is now seeking donations to buy stock in television companies so it can object formally to trash-science. Shares can be purchased also by individuals, and even a single share provides the holder influence with the company.

11. *Write letters to the editor and op-ed commentary* for newspapers and magazines expressing your views on pseudoscientific programs, events, books, and articles.

12. *Actively support high quality science education from grade school through continuing adult education.* Concerned citizens must insist that local, state, and federal governments fund science education fully. Some school boards take a decidedly antiscience approach and even prohibit the teaching of valid science (for instance, of biological evolution). These must be watched carefully and countered by the scientific community. And special effort should be taken to ensure that teacher training programs in colleges and universities include real science, backed where possible by educational programs provided by professional scientific societies and organizations.

References:

Two kinds of references are listed below: Standard print media available in most larger libraries, and Internet references given by their World Wide Web location. The addresses of some of the WWW sites may change in time, or that listed for a particular document may be difficult or impossible to access. If so, go first to the general site location (in brackets, listed between the // and /), then examine the information there for the document listed below. Usually, a search on the title or subject in one of the WWW search engines (such as Yahoo, AltaVista, or HotBot) will produce the cited and related documents.

Blake, W.P. 1899. The Pliocene skull of California and the flint implements of Table Mountain. *Journal of Geology* **7**: 631- 637.

Corso, P.J. 1997. *The Day After Roswell* (New York: Pocket Books), 341 pp.

Ehrlich, P.R. and Ehrlich, A.H. 1996. *Betrayal of Science and Reason: How Anti-Environmental Rhetoric Threatens Our Future* (Washington, D.C., Island Press), 335 pp.

Emery, C.E. 1997. How do TV's pseudoscience specials rate? And the case of the Quadro Tracker. *Skeptical Inquirer* **21:** 20-21.

Eve, R.A., and Harrold, F.B. 1991. *The Creationist Movement In Modern America* (Boston: Twayne), 234 pp.

Fawcett, L. and Greenwood, B.J. 1992. *The UFO Cover-Up. What the Government Won't Say* (New York: Fireside Books, Simon and Shuster), 259 pp.

Hastings, R.J. 1987. New observations on Paluxy tracks confirm their dinosaurian origin. *Journal of Geological Education* **35**: 4-15.

Hirshon, R.D. and Lederman, L.M.. 1996. Planning Grant for a "Prime-Time" TV Science Drama. (Washington, D.C.: National Science Foundation). [http://www.nsf.gov/cgi-bin/showaward?award=9550660]

Kuban, G.J. 1989. A matter of degree: An examination of Carl Baugh's alleged credentials. *National Center for Science Education Reports* **9**(6): 15-18. [Also at: http://members.aol.com/Paluxy2/degrees.htm]

Leroux, H., Warme, J.E., and Doukhan, J.C. 1995. Shocked quartz in the Alamo Breccia, Southern Nevada—evidence for a Devonian impact event. *Geology* **23**: 1003-1006.

Lederman, L.M. 1996. A strategy for saving science. *Skeptical Inquirer* **20**: 23-28.

Lett, J. 1990. A field guide to critical thinking. *Skeptical Inquirer* **14**: 153-160.

Limbaugh, R. 1992. *The Way Things Ought to Be* (New York: Pocket Books), 304 pp.

Limbaugh, R. 1993. *See I Told You So* (New York: Pocket Books), 364 pp.

Marrs, J. 1997. *Alien Agenda* (New York: Harper Collins), 320 pp.

McDevitt, T.M. 1996. *World Population Profile: 1996* (Washington, D.C.: U.S. Census Bureau). [http://www.census.gov/ipc/www/wp96.html]

Miller, J.D. 1987. The scientifically illiterate. *American Demographics* **9**: 26-31.

National Science Board. 1996a. *Science and Engineering Indicators—1996. NSB 96-21* (Washington, DC: U.S. Government Printing Office). [http://www.nsf.gov/sbe/srs/seind96/start.htm]

National Science Board. 1996b. *US Science and Engineering in a Changing World. NSB 96* (Washington, DC: U.S. Government Printing Office), 4 pp. [http://www.nsf.gov/sbe/srs/seind96/start.htm]

Perkins, R., Jr. 1995. *Logic and Mr. Limbaugh* (Chicago: Open Court), 182 pp.

Sacks, P. 1996. *Generation X Goes to College: An Eye-Opening Account of Teaching in Postmodern America* (Chicago: Open Court), 208 pp.

Sagan, C. 1996. *The Demon-Haunted World—Science as a Candle in the Dark* (New York: Random House), 457 pp.

Scott, E.C. and Cole, H.P. 1985. The elusive scientific basis of creation "science." *The Quarterly Review of Biology* **60**: 21- 30.

Shermer, M. 1997. *Why People Believe Weird Things: Pseudoscience, Superstition, and Other Confusions* (New York: W. H. Freeman), 306 pp.

US Census Bureau. 1997. Population topics. [http://www.census.gov/ population/www/]

Wade, C. and Travis, C. 1990. Thinking critically and creatively. *Skeptical Inquirer* **14**: 372-377.

Warme, J.E. and Sandberg, C.A. 1996. Alamo megabreccia: Record of a Late Devonian impact in southern Nevada. *GSA Today* **6**: 1-7.

Wheeler, M. 1996. Yes, we have no yellow fruit. *Discover* **17**(11): 118 (6 pp.).

CHAPTER 5

BREAKTHROUGH DISCOVERIES

J. WILLIAM SCHOPF[1]

INTRODUCTION

From time to time mind-boggling scientific breakthroughs are trumpeted in the press—chemicals to control AIDS, fossils of ancient humans, the cloning of Dolly the sheep, insights into the beginnings of the Universe. Because there are more scientists than ever before, breakthroughs come at a quickening pace. Yet truly major ones still are few and far between, precious rarities worth savoring.

The goal in science always of course is to "get it right." But getting a wrong answer matters more some times than others. Science would not be set on its ear if a fossil animal said to be "new to science" were years later shown identical to one of scores of similar species from the same rock layer. But it would be a disaster if science accepted the "breakthrough discovery" of an animal wrongly reported from a bed a billion years in age, hundreds of millions of years older than any other ever found. The first error would be minor, unlikely to have much effect on established knowledge. The second, serious because its acceptance would tear down a corner-stone of human understanding.

Because a claim of a breakthrough is just that—announcement of a find said to be surpassingly important—those making the claim have special responsibilities. The late Carl Sagan was right—extraordinary claims *do* require extraordinary evidence.

Extraordinary claims are as old as science itself. Those shown correct are well chronicled (especially when they lead to Nobel Prizes) but even more fascinating are failed claims because they reveal a side of science rarely seen, of error, despair, sometimes even of eventual redemption of the claimant if not the claim. Two of the most amazing from the annals of paleontology are recounted here—salamander fossils mistaken for human skeletons buried by Noah's Flood (!) and fanciful fake fossils hand-carved to mystify and embarrass an arrogant professor. The chapter concludes with a first-hand account of events surrounding another extraordinary claim, the recent report of evidence of ancient life on Mars.

Though marvelously entertaining, the true value of these tales is what they teach about the workings of science, past and present.

[1]IGPP Center for the Study of Evolution & the Origin of Life, Department of Earth & Space Sciences, and Molecular Biology Institute, University of California, Los Angeles, CA 90095.

"MAN, A WITNESS OF THE DELUGE"

One of the most famous extraordinary claims in paleontology comes from the early 18th century and the studies of Dr. Johann Jacob Scheuchzer (1672–1733), a Swiss physician and naturalist of exceptionally broad-ranging interests. Best known for his studies of geology and paleontology, he was a respected scholar also of mathematics, geodesy, geography, literature, and numismatics (coin collecting). An outstanding academic, the *English Cylopedia* of 1856, more than a century after his death, acclaims him as a savant of "indefatigable industry and extensive knowledge."

Scheuchzer was well-schooled in the Christian tradition. Like many learned men of the day a confirmed **Diluvialist**, his view of Earth history was molded mightily by the story of Noah and the Flood. In his 1709 treatise *Herbarium Diluvianum*, for instance, he not only pegs the Great Deluge as the forming agent of fossil-bearing rocks but actually dates the Flood as happening in the spring, probably in May, because of the "tender, young, vernal" state of entombed seed cones. (This notion—likely borrowed from Englishman John Woodward's 1695 *Essay towards a Natural History of the Earth*—was hotly disputed by those who argued that "ripe" fruits in the very same deposits proved the Deluge happened in autumn.)

But Johann Jacob Scheuchzer is most remembered for a later contribution, a discovery that was to be the center of debate for nearly 100 years.

In 1725 Scheuchzer uncovered the partial skeleton of a large, elongate, obviously vertebrate animal (Figure 5.1) in limestone quarried near Oeningen, Baden, Germany. To Scheuchzer the preserved skull and backbone looked decidedly humanlike and totally different from any fossil he had ever seen. There could be only one explanation—this must be the remains of a man drowned in the Flood, the bones of one of those miscreants whose sinful ways brought upon the world the catastrophe of forty days and nights of the Great Deluge!

Elated with this proof of Biblical Truth, Scheuchzer made it the keystone of his monumental *Physica Sacra* published five years later. In his words:

> It is certain that this ... is the half, or nearly so, of the skeleton of a man: that the substance even of the bones, and, what is more, of the flesh and of parts still softer than the flesh, are there incorporated in the stone. We see there the remains of the brain ... of the roots of the nose ... and some vestiges of the liver. In a word it is one of the rarest relics which we have of that cursed race which was buried under the waters [of the Noachian flood].

To memorialize the find Scheuchzer dubbed the specimen *Homo diluvii testis*—literally, "Man, a witness of the Deluge"—and for dramatic effect included a moralistic couplet penned by church deacon Miller von Leipheim:

> Betrübtes Beingerüst von einem alten Sünder,
> Erweiche, Stein, das Herz der neuen Bosheitskinder.

In English (as translated by Herbert Wendt):

> Afflicted skeleton of old, doomed to damnation,
> Soften, thou stone, the heart of this wicked generation!

Within only a few years Scheuchzer's antediluvial man was hailed throughout Christendom as irrefutable evidence of the Holy Word.

But not all scholars were convinced. Among the first to question Scheuchzer's

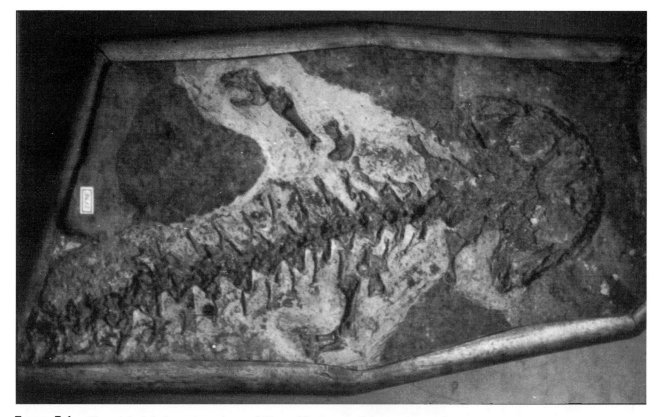

FIGURE 5.1   The original (holotype) specimen of *Homo diluvii testis*, "Man a witness of the Deluge," discovered in 1725 near Oeningen, Germany, by Johann Jacob Scheuchzer and thought by him to be the flattened skull and backbone of a man drowned in Noah's Flood. Known now to be the upper torso of a giant salamander fossilized in 15 milion year-old (Late Miocene) freshwater limestone, the skeleton was hailed for nearly 100 years as direct proof of Biblical Truth. (Teyler Museum specimen 8432, Case V29 in the second room of the Minerals and Fossils Cabinet.)

claim was another Swiss physician and naturalist, Johannes Gessner (1709–1790), a former student of Scheuchzer and his successor to the Professorship at the Carolinum in Zürich, Switzerland. Gessner raised many an eyebrow by suggesting in 1758 that Scheuchzer's famed find might actually be bones of a large though only partially preserved fossil fish.

Doubts of Gessner and a few others not withstanding, to most the matter was settled. The Bible was beyond question. Scheuchzer's view was eminently sensible, confirmation of what conventional wisdom already knew.

At his death in 1733 Schuchzer bequeathed the specimen to his son who passed it along to his son. In 1802 and now world-renowned, *Homo diluvii testis* was purchased from Schuchzer's grandson by the Teyler Museum (Figure 5.2) at Haarlem, the Netherlands, just west of Amsterdam, where it is housed to this day.

Cuvier "Cleans" the Specimen

A few years later the famous fossil came under the scruity of Baron Georges Cuvier

(1769–1832), the French comparative anatomist credited as a founder of the science of paleontology.

In 1810, with annexation of North Germany and the entire kingdom of Holland, Napoleon Bonaparte's empire reached its widest extension. Cuvier was Napoleon's Minister of Education and in 1811 journeyed to Amsterdam as head of a commission to review and improve the Dutch educational system. He had written ahead to the Directors of the Teyler Foundation requesting permission to examine and "clean" the specimen, to chip away rock that masked parts of the skeleton from Scheuchzer's view. *Homo diluvii testis* was far and away the most precious specimen of the museum's collections—kept under lock and key and cleaned not even by Scheuchzer when it was unearthed in the 1720s—but to curry favor with the occupying French, the Directors granted Cuvier's request.

With painstaking skill Cuvier uncovered the fossil's "arms" and "hands." Rather than human, as Scheuchzer surmised, these turned out to be the short forelimbs and clawed forefeet of a salamanderlike amphibian! Cuvier was not

Figure 5.2 Teyler Museum, in Haarlem on the outskirts of Amsterdam, built in 1780 behind the home of the museum's benefactor, Pieter Teyler van der Hulst, and the oldest museum built as such in The Netherlands. A childless widower and owner of a prosperous silk-mill, at his death in 1778 Teyler bequeathed his house and much of his fortune to establish a foundation bearing his name for "promoting [the Christian] Religion, encouraging the Arts and Sciences, and for the Public Benefit." The museum consists of four "cabinets" (departments), of Art, Coins, Physics, and Minerals and Fossils.

surprised. On the basis of Scheuchzer's published drawing (Figure 5.3, left) Cuvier had long suspected the fossil to be a large salamander, and guided by a sketch of a salamander skeleton brought with him from France he chipped away just those parts of the rock needed to prove his point.

If one compares the actual specimen (Figure 5.1) with the drawings of Scheuchzer and Cuvier (Figure 5.3) it is easy even today to discern Cuvier's hand, to see exactly where he "cleaned" the storied skeleton so many years ago.

Cuvier showed *Homo diluvii testis* to be a fossil salamander belonging to the genus *Andrias* so in 1831 it was renamed *Andrias scheuchzeri*, in honor of Scheuchzer. But the salamander's large size remained a puzzle. Its upper torso, the part preserved in Scheuchzer's specimen, is nearly half a meter long so the animal would have been more than a meter from nose to tail, much larger than any salamander ever seen.

This quandary, too, was soon resolved. In 1829, nearly two decades after Cuvier's restudy of the specimen and more than a century after it was found, the German explorer Philipp Franz von Siebold (1796–1866) discovered giant batrachian salamanders living in southern islands of the Japanese archipelago, modern forms so similar to the fossil in size and bone structure they have been named a subspecies, *Andrias scheuchzeri japonicus*.

In the century and a half since, thousands of specimens of fossil plants and

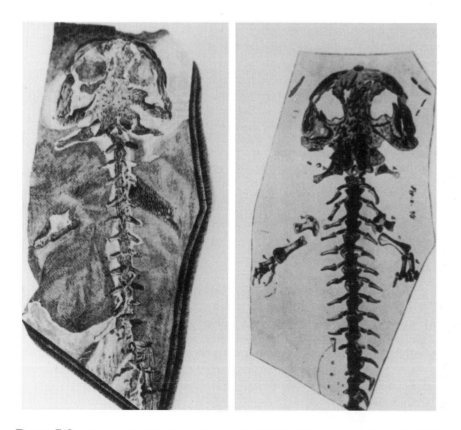

FIGURE 5.3 *Homo diluvii testis*, as illustrated in 1735 by Scheuchzer (left) and in 1824 by Cuvier (right) after he had "cleaned" the fossil (renamed *Andrias scheuchzeri* in 1831) to expose its forelimbs and clawed forefeet that showed it to be a salamander, not as Scheuchzer thought a human skeleton buried in the Biblical Flood.

animals, including hundreds of different species, have been unearthed from the freshwater limestone at the Oeningen quarry, known now to be Late Miocene (about 15 million years) in age. Though rare, giant salamanders are known from some 26 examples, several nearly a meter-and-a-half long.

For nearly 100 years after its discovery in the early 1700s *Homo diluvii testis* was acclaimed as proof of Noah's Flood. But it was no proof at all, simply a misinterpreted fossil salamander, unusually large and new to science. Known now as *Andrias scheuchzeri*, the name of Johann Jacob Scheuchzer is forever linked to his mistakenly identified "Man, a witness of the Deluge."

BERINGER'S LYING STONES

The legend of "The Lying Stones of Dr. Beringer" tells the story of yet another extraordinary paleontologic claim. Told, retold, and embellished in paleontology courses worldwide for more than two-and-a-half centuries, it is an incredible story of an infamous hoax that provides a rare peek into the workings of natural science in the early 1700s.

The standard story is that of an imperious professor, J. B. A. Beringer, duped by his students who carved and hid fake fossils where the prof was sure to find them. He did, thought he had chanced on a gold mine, and with grand flair wrote up the discovery in a magnificent opus. But he then found is own name spelled out on one of the rocks in Hebrew letters. He'd been hoaxed! Humiliated, he bought back the volumes and died soon thereafter, penniless, friendless, in deep despair.

A wonderful fable, but only partly true. Some years ago the real story came to light thanks to 200 year-old court documents discovered in the Würzburg State Archives.

• In 1726, Johann Bartholomew Adam Beringer (1667–1740) *did* author the grand opus (the *Lithographiae Wirceburgensis*), his third published work. But he

Figure 5.4   One of the surviving specimens of Beringer's famous lying stones showing a bird in flight. (From the collections of the Geologisch-Paläontologisches Institut der Universität Würzburg, Germany.)

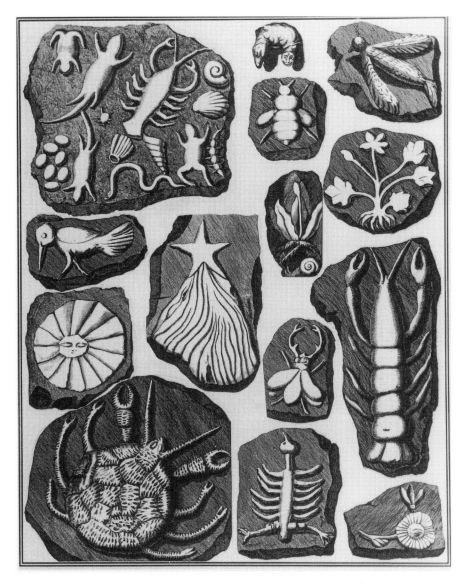

FIGURE 5.5 Representative figures from the *Lithographiae Wirceburgensis*. Among the various fantastic "oryctics" (things dug from the ground), note the bony bird (lower center); a flower being visited by a bee (lower right); a smiling Sun (left center) and firey comet (center); a plant with root, stem, leaves, and underlying snail (right center); a feathered bird in flight (upper right); a smiling two-headed lizardlike oddity (upper right center); and a diverse assemblage of "fossils" (upper left).

did not die until 14 years later and during the interim published two more volumes, one a lengthy treatise on the spread and treatment of the dreaded plague.

• Beringer *was* a professor at the University of Würzburg, Germany. A physician, son of a professor, and recipient of Ph.D. and M.D. degrees, he held the titles of Senior Professor and Dean of the Faculty of Medicine at the University, Chief Physician at Würzburg's Julian Hospital, and Advisor and Chief Physician to Christopher Franz, Prince-Bishop of Würzburg and Duke of Franconia.

• His treatise *does* depict an astounding zoo of fake fossils (Figures 5.4 to 5.6). Butterflies, beetles, bees (one together with its honey-comb). Birds (some

Figure 5.6 Representative figures from the *Lithographiae Wirceburgensis*. Particularly notable are the "perfectly preserved" millipede (lowermost right); a spider with its web (lower right center); a mermaidlike sea creature (lower right); a four-toed lizardlike form (center right); a spider devouring a fly (uppermost left); ants and wasps with their nests (upper left center); and copulating frogs (lower center), salamanders (uppermost right), and insects (center and upper left). The fabricators of Beringer's famous "Lügensteine" (lying stones) must have had a sense of humor!

feathered, others denuded, one in flight, two alongside freshly laid eggs). Flowers (with their stems, roots, and leaves, one being pollinated by a hovering bee). Fish, frogs, salamanders, lobsters. Crabs, millipedes, scorpions, earthworms. Ants and wasps with their nests, spiders (two with their webs, one devouring a fly). Caricatures even of the Moon, Sun, stars, and comets. Many stones still exist

(Figure 5.4), some on view at the Geologisch-Paläontologisches Institut der Universität Würzburg, others at natural history museums in Berlin and London.

• The famous "Lügensteine" (literally, lying stones) *had* been carved, hidden (on a low hilltop, Mount Eivelstadt, on the outskirts of Würzburg), then "dug up" as part of an elaborate hoax. But the finds were made by diggers employed by Beringer, not by him, and the perpetrators were not Beringer's students. Instead the scam was pulled off by two academics, Herr J. Ignatz Roderick (Professor of Geography, Algebra and Analysis) and the Honorable [sic] Georg von Eckhart (Privy Councilor and Librarian to the Court and to the University), aided by a German nobelman named Baron von Hof (possibly the backroom financier of the conspiracy but about whom little is known except that he was trundled about in a sedan chair).

Soon after the *Würzburg Lithography* was published, Beringer came to realize the hoax, and when the perpetrators then spread rumors accusing *him* of the fraud he requested a special judicial hearing held in April 1726 for "the saving of his honor." He was exonerated by the testimony of Christian Zänger, one of Beringer's teenage diggers, who confessed his complicity and fingered Roderick, von Eckhart, and the Baron whom he overheard hatch the plot to ruin Beringer "because he was so arrogant and despised them all." Von Eckhart died a few years later and the Baron escaped the scene unscathed. But Herr Roderick, the ringleader of the plot, absented himself from Würzburg, evidently banished from from Duchy of Franconia

• Beringer *did* discover Hebraic lettering among his "fossils." But the discovery was made before, not after, publication of his opus; the writing was in Latin and Arabic, not only Hebrew; and the letters spelled out "Jehovah" instead of

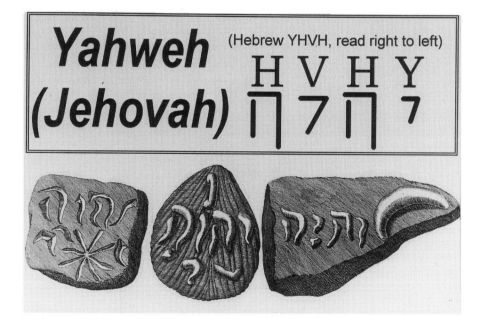

FIGURE 5.7 Three of the stones depicted in Beringer's *Lithographiae Wirceburgensis* that show the surface marks (accompanied by ornamentation) he interpreted correctly as Hebraic lettering for "YHVH"—Yahweh (Jehovah). Other stones said by Beringer to show the name of God in Latin and Arabic characters are not pictured in the *Würzburg Lithograpy.*

Beringer's name (Figure 5.7).

• And original copies of Beringer's treatise *are* of almost legendary rarity, greatly prized by bibliophiles. A spurious "second" (title-page) version, issued in 1767 evidently at the behest of Beringer's heirs, is not nearly so scarce. But only a few copies of the 1726 first edition have been available for sale in recent decades, exclusively at booksellers' auctions (where their price skyrocketed from $125 in 1961, to $250 in 1971, to more than $1,000 in 1987).

It is no surprise that Beringer and his book have long been remembered. His seems the story of an overzealous, imperious, and incredibly naive "savant" who was unwittingly duped (by trusted colleagues, contemptuous of his arrogance) into accepting carved stones as fossils and thereby making a fool of himself in the eyes of the world. The moral of the saga seems inescapable.

Yet Beringer assuredly was no bumpkin. He held high positions both in the University and the Court of Würzburg; was praised in his lifetime as an "illustrious ... most inquisitive scholar"; and even in the 1880s, 150 years after the fiasco, acclaimed in the official history of the university as a "virtuoso ... tireless scholar ... the most active man of his time." What led this touted academic to ruin? His stones bear only the most superficial resemblance to true fossils and it is hard to imagine that anyone could be fooled by "fossilized" stars, moons, comets, and Hebrew (or Latin or Arabic) letters!

To fathom Beringer's blunder we first need to understand what was known and what was not about the nature of fossils. His was a time of flux in paleontology, of competing ideas and uncertain answers. There was even question whether "lithology" (as the study of rocks, minerals, and fossils then was known) was a pursuit worthy of learned men, a question raised by Beringer in Chapter I of his remarkable tome (and rendered from his stylized flowery Latin by M. E. Jahn and D. J. Woolf in *The Lying Stones of Dr. Beringer*, a classic published in 1963 by the University of California Press, Berkeley):

> To what purpose, they ask, do we stare fixedly with eye and mind at small stones and figured rocks, at little images of animals or plants, the rubbish of mountain and stream, found by chance amid the muck and sand of land and sea? To what purpose do we, at the cost of much gold and labor, examine these findings, describe them in vast tomes, commit them to engravings and circulate them about the world, and fill thick volumes with useless arguments about them? What a waste of time and of the labors of gifted men to dissipate their talents by ensnaring them in this sort of game and vain sport! Does this not amount to neglecting the cares of the realm to catch flies ... the efforts, the genius, and the expenses of learned men gone mad?

Beringer's answer shows a devotion to knowledge motivated by strong religious underpinnings:

> The mind of man, [however], was made for higher things, not for the fattening of the belly, nor for the luxury and delights of the body, nor for that most inane occupation of all, the custody of a pile of gold or brass. ... We do not recognize as scholars of the fine arts those who are the slaves of lucre ... but rather those noble souls who expend their energies ... stimulated by the very dignity of learning and of laudable work, in order to experience the pure joy born of knowledge. This is the core of Christian ethics. ... What more noble purpose for human actions can be conceived than that whereby from the marvelous effects of nature we ascend ... to the recognition of the power of the Creator?

Cabinets of Curiosities

In Beringer's day the study of fossils was carried out mostly by physicians, almost always as a hobby when duties and time permitted. Many of these learned dilettantes assembled "**cabinets**," private collections of natural rarities—rocks, minerals, shells, feathers, animal skins, archaeological oddities. Those with especially noteworthy cabinets often bequeathed them to local municipalities, collections that even today can be ferreted out among the holdings of major museums.

There is no evidence that Beringer's cabinet—"gathered on nearly all the shores of Europe" and consisting mostly of "oryctics," things dug from the ground—was in any way remarkable prior to May 31, 1725. But on that day his diggers found and delivered to him the first three figured stones, one bearing a "circle, like the sun with its rays," the others wormlike shapes. Within a scant six months hundreds of stones, some having figures on both sides, were added to his collections. Beringer's cabinet had come to be remarkable, indeed!

Was Beringer pleased? Thrilled? Ecstatic? You bet! From the Introduction to *Lithographiae Wirceburgensis:*

> These wonderful exhibits reveal the most bountiful treasure of stones in all of Germany ... at last, thanks to my persevering effort ... discovered and unearthed at no small cost and labor. I doubt that any more heart-warming spectacle can come to the eye of a scholar of natural science.

But he was also deeply perplexed, at a loss to make sense of his "wonderful exhibits." So he did what any scholar would do, he trekked off to the library and looked up relevant works to see whether anyone else had come upon stones like his and a way to explain them.

Beringer Begs the Question

Only a few scholars of the time knew as we do today that fossils are the bodies or impressions of organisms preserved naturally in rock. Of these probably the best known is Robert Hooke (1635–1703), first Curator of Experiments to the Royal Society of London and remembered in modern textbooks as a pioneering microscopist.

But Hooke's view, far from prevalent in the 1700s, is largely ignored by Beringer who deals chiefly with three much more popular notions:

1. The theory that fossils are so-called sports of nature, stones formed by an extraordinary "plastic power" present in the Earth.
2. The idea that that fossils grew in place from the seeds and eggs (the "Spermatic Principle") of plants and animals lodged in crevices by rain or ground water and activated by heat, saline moisture, or some other supposedly life-giving force.
3. The concept that fossils are remains of organisms buried during the Biblical Flood.

Because none of the theories seemed to fit his finds, Beringer considered less popular notions as well—that the figures on his stones were formed by the influence of the stars (essentially an astrological explanation), were impressed on them by the "plastic force" of light, or were the handwork of prehistoric pagan tribes. But neither these nor the more conventional views meshed with the evidence at hand: "I attempt to demonstrate...that the nature of our stones is so unusual that its very novelty eludes the [above noted] opinions, however well established by documentation and searching experimentation [they may appear to be]."

Until the fraud finally came to public notice Beringer remained convinced his stones were "works of Nature." But unable to explain how nature made them he ends his treatise by begging the question, arguing that a final judgment is best left to "wise men…to the men of letters, to the scholars and patrons of the finer and more profound disciplines" who in time would render the ultimate verdict on the significance of his beloved "earthly treasures."

Unearthing a Rosetta Stone

In 1725 Beringer thought he had discovered a Rosetta Stone, though he couldn't make out its meaning. In the very same year Scheuchzer was *certain* he had unearthed a Rosetta Stone, one linking the Biblical past to the present. Beringer may have been naive to be so fooled, but was smart enough not to claim more for his find than he could fathom. Scheuchzer shoehorned his to fit accepted dogma and misled science for nearly 100 years. In the 20/20 hindsight of two-and-a-half-centuries both are seen to have missed the mark.

Yet who are we to sit in judgment? Like Beringer and Scheuchzer we carry out our daily work saddled with ideas we have been taught are true and struggle just as they did to make lasting contributions. But some of what passes today as "known" is sure to turn to dust. And though we guard against errors, almost no scientist is immune from the heady pull of a major breakthrough, only few from the urge to shoehorn finds to fit accepted views. The most we can ask is to do the best we can—a best that history shows inexorably *does* get better!

THE HUNT FOR LIFE ON MARS

The news rocketed around the world: "PAST LIFE FOUND ON MARS!" Asked by NASA to give the public a first-blush scientific appraisal of this extraordinary claim, I was on hand at the Washington, DC news conference that announced the find in August 1996. But my involvement dates from earlier.

At the request of NASA administrators, in January 1995 I journeyed to the Johnson Spacecraft Center (JSC) in Houston, Texas, to render a first-hand opinion on what geologist-mineralogists there believed might be microfossils in a chunk of meteorite thought to come from Mars. The JSC scientists swore me to secrecy lest their find hit the newspapers before they had the facts.

What caused the fuss were tiny orange-colored pancake-shaped globules of **carbonate** ($CO_3^=$-containing) mineral, 2 to 200 µm across and ringed by thin black and white rinds. Flushed with excitement, the researchers explained that never before had ringed discs like these been seen in a meteorite, and since this one was said to have come from Mars—which once may have harbored life—and since the objects were made of the same mineral and some about the size as shells of certain earthly protozoans (foraminiferans), they thought they might have chanced on a mélange of Martian fossils.

Because of my studies of the most ancient (nearly 3.5 billion year-old) fossils on Earth, I had been brought in to shore up the paleontologic guess of scientists schooled in rocks and minerals, but not biology.

Their guess was wrong. A number of the objects were simple discs but many merged one into another in a totally nonbiologic way. Their size-range also did not fit biology and they lacked any of the telltale features—pores, tubules, wall layers, spines, internal structures—that earmark tiny shells. Moreover, the "lifelike" traits they did possess (carbonate composition, discoidal shape, ringed rims) could be explained by ordinary inorganic processes.

Carbonate minerals are, of course, laid down by life (not only in the shells of protozoans but those of clams, snails, and other animals), but they also form by purely inorganic means and are known from many meteorites, not just the one containing the putative fossils, where their nonbiologic genesis is beyond question.

The shape of the objects didn't seem to need biology either. Formed when mineral-bearing solutions percolated through a thin crack in the rock, the pancakes are flat on top and bottom because the solidifying carbonate ran out of space above and below. And they are more or less circular because as minerals drop out of solution they crystallize around all sides of the grain first formed (the "center of nucleation"), in this case making a disc.

Their rimming rings, I thought, came from the same process. When the makeup of a crystallizing solution changes so do the minerals laid down. The thinly layered black and white rims showed that chemical conditions changed as the pancakes formed, not that the discs were formed by protozoans.

I raised these points with the JSC scientists. They seemed to agree. The matter, I thought, was closed.

But I urged them to continue the hunt. I believed then as I do now that the search for hints of life in Martian meteorites is a promising way to attack a truly fascinating question. It *is* important to know whether life once existed (or still does) on Mars. (Still, I was taken aback at the August 1996 news conference when the same little pancakes were again proffered as evidence of Martian life, this time of bacteria rather than "protozoans." The facts hadn't changed, only the meaning attached to them.)

NASA STAGES A PRESS CONFERENCE

Several weeks before the August news conference I received a phone call from NASA Headquarters informing me that the scientists I had visited more than a year earlier in Houston had completed studies claimed to provide evidence of ancient life on Mars. A technical article reporting their results was soon to appear in *Science*, a highly regarded journal reserved for the hottest of hot discoveries. NASA felt obliged to inform the public and planned to do so at a pre-publication press conference. But because some at Headquarters thought the evidence "a bit iffy" they wanted an outside expert to appraise the findings publicly when they were announced to the world. Would I, please, perform this task?

I was reluctant. I had plenty on my plate already and feared this was one more in a string of spurious claims for "life in meteorites" that dates to the early 1960s. Still, I hadn't read the article, hadn't seen the evidence. And the scientists making the claim were colleagues. I agreed "to think about it."

A copy of the soon-to-be-published report arrived the next day. I studied it. Carefully. Three times. I was not impressed. Though some of the report was backed by solid scientific data, support for other parts was wanting. Crucial questions had not been asked. Works published earlier and critically relevant to the authors' discussion had been ignored. Alternative, to me more plausible ways to explain the findings were given short shrift. The manuscript's concluding claim of "evidence for primitive life on early Mars" seemed overblown, ill-conceived.

I called NASA, and quoting Carl Sagan's catchphrase that "extraordinary claims require extraordinary evidence" opined that for this claim the evidence was not even close to that required. I suggested names of three other scientists to serve in my stead.

A few days later NASA called back. NASA Director Dan Goldin had personally pegged me for the job, partly, I gather, because he's a Sagan fan (and was said to have been pleased by the quote), but I think mostly because he knows it's in NASA's best interest to get the story straight. Any claim of life on Mars—whether of organisms small or large, past or present—is bound to stir controversy. This one would be no exception. The "iffy" evidence was certain to raise eyebrows and since NASA's budget hearings were looming, even the timing of the announcement might be regarded as suspiciously fortuitous. My guess is that Mr. Goldin—a truly able administrator and brilliant politician (appointed by Republican Bush, a star of Democrat Clinton's team)—figured a preemptive strike was in order. To protect NASA's scientific integrity and the same time stiffle the easily predicted army of naysayers, he decided to assign a hard-nosed outsider to evaluate the claim. Who better than one calling for it to be backed by "extraordinary evidence?"

Before the call relaying Goldin's personal request I thought I was in the clear. This was a task I did not want to do. But Goldin is the NASA boss—the "faster, cheaper, better" guy—an appointee of two presidents. Who am I to turn him down?! I agreed.

Prelude to the Feeding Frenzy

The news conference wasn't scheduled for another two-and-a-half weeks. I tried to put it out of my mind. But by the next weekend I'd become increasingly concerned. My skepticism was bound to raise some hackles.

I spent a couple of days listing my arguments on vu-graphs (see-through charts like those NASA often uses) and early on the following Tuesday Faxed copies to Houston. It was only fair to warn the JSC group what I planned to say. But I also wanted to make certain I had not misunderstood the technical details of their article. I was sure they'd straighten me out.

My hope for dialogue came to nothing. Neither they nor I had time. About an hour-and-a-half after I sent the Fax I received yet another call from Washington: "Bill, get on the 1:30 afternoon flight. The press conference has been moved up."

I arrived at Dulles Airport late that night and at NASA Headquarters the next morning where I was squired to a basement room in which I found the JSC team practicing its lines. They were prepared. Thoroughly. They even had a high-tech cartoon-video to tell the story of the flight of the meteorite from Mars through space to us. And though the room lacked a VCR to show the video, they didn't miss a beat. When they came to that part of their run-though one of the team said: "My video talkover lasts 2 minutes, 47 seconds." The one next to him laughed: "Mine's only 2 minutes, 19." (VCR-blind, the first gave his rendition practically the same as he gave it later to the reporters upstairs. The 2' 19" version changed not at all. These folks were pros!)

Finally my turn came in the practice session. They had videos. I had vu-graphs. They'd practiced. I'd not. They were NASA. I, an outsider.

I gave my spiel. By that time there was a pride of NASAites overlooking our run-through, Dan Goldin included. I finished. Utter silence. Then a woman on the Headquarters' staff rose and berated the troops: "Schopf has just demolished you. Can't you guys be more positive?!" (I don't know who this person was—was never introduced, never caught her name—but you can see her on the CNN tape of the press conference introducing Administrator Goldin). The JSC crew was in a quandary. Like I, they knew their story was circumstantial. There was no "smoking gun." Yet it was important for them to look good, to please the boss. The pressure was great. They seemed torn.

At the practice session I tried to be reasonable, even gentle. I did, too, at the later news conference, a performance for which I've been much praised—but also chastised (by no less than a Nobel Laureate!), for being too soft. Still, it seems to me that the "Mars Science Team" (as they were now calling themselves, bolstered by input from scientists at McGill, Georgia, and Stanford Universities) tackled a difficult interdisciplinary problem. An instant answer, pro or con, was not in the cards.

Breaking the News to the World

Not only had I not practiced for the news conference, I not been warned what to expect. Maybe no one knew. The only thing I had to go on were memories of the late '60s when I and five other scientists (officially, the Lunar Sample Preliminary Examination Team) were tasked to do the first studies of Moon samples gathered on the Apollo 11 and 12 missions. While the Apollo crews rested in quarantine in another part of the building, we sorted, studied, and described the rocks. To test whether they harbored virulent Moon-germs (dubbed "Gorgo" by us), we even monitored the effects of lunar dust fed to Japanese quail, germ-free mice, and various plants (some of which grew better than on Earth soil). Interactions with the media were friendly, interviews one-on-one or at most with a few pool reporters from magazines, newspapers, radio, TV.

The Mars news conference could not have been more different. Instead of a few reporters there were 500. Instead of note pads there were scores of video cameras. There was so much electronic gear in the auditorium that the sound system overloaded and the conference had to be delayed to take care of feedback whining through the hall.

On the stage I was seated alongside the chief of the Stanford group that identified organic compounds in the meteorite. Just before the conference was to begin he waved to a friend among the gaggle of journalists. Within only a few seconds he, and I next to him were besieged by a churning sea of microphone-thrusting reporters, all determined to shove to the front of the pack. A media feeding frenzy!

Things quieted down and we waited for another 20 minutes as coverage switched to the south lawn of the White House where President Clinton read a carefully crafted statement on the significance of the find about to be revealed. For the next two-and-a-half-hours CNN carried our press conference to the world. Mr. Goldin led off, followed by the well-choreographed presentations of the science team. My remarks came last, succeeded by a lengthy session of questions from the Washington press corps and journalists assembled at NASA installations across the country and answers from Mr. Goldin and those of us at the dais.

The researchers' presentations were measured, sensible, their arguments plausible. By the time they finished I think everyone was willing to believe. Introduced as the designated "skeptic" to "begin the debate," I had no doubt my words would prove unwelcome. But I had no choice. The evidence was (and is still) inconclusive, and it fell on me to point that out. Some claim the glass is half full. To others it's half-empty. But no one who knows the facts would claim it's overflow-ing—not then, not now, not even the Mars Team.

METEORITES FROM MARS

Mars As an Abode for Life

Notions of life on Mars date from the early days of modern science. And for good reason—the Red Planet is a cool place! It has ancient rusty plains pockmarked like the Moon, and the longest (4,000 km) and deepest (nearly 10 km) canyon—Valles Marineris—known anywhere. It boasts the largest volcano ever seen (Olympus Mons), some three times taller than Mt. Everest, and sinuous channels carved by rivers now long-dry. It never rains on Mars and its mostly (95%) carbon dioxide atmosphere is so thin, one-tenth the pressure of Earth's, that without the cocoon of a protecting spacesuit one's eyes would pop out.

But in other ways Mars is like a smaller brother to our planet, a rocky body half the size with one-third the gravity but with a day only 37 minutes longer and seasonal swings (in a 669-day year) much like Earth's. At Carl Sagan Station, where NASA's Pathfinder landed in July 1997, temperatures range during the Martian summer from that of a freezing winter day in southern Canada to the coldest on Earth.

Yet not always was Mars so cold and dry. Though never close to tropical, its youthful climate was much more hospitable to life like ours. This is one of the keys to the past-life-on-Mars story. The Martian meteorite dates from early in Mars' history, an epoch when rivers flowed, the atmosphere was thicker, and life may have gained a foothold.

A second key is that the story centers on minute forms of life, bacteria, not large organisms such as ourselves. We now know that simple single-celled microbes play a far larger role in the evolutionary Tree of Life and are much more resilient than previously thought. They exist on Earth in a strikingly wide range of conditions—in boiling deep-sea vents, sulfurous acid springs, cracks and crevices in rocks deep in the crust, on and within ice sheets and permanently frozen Arctic tundra, even in mineral-encrusted fissures in the rocks of Marslike ice-cold deserts. If they can survive, even thrive, here, why not also on Mars?

ALH84001

The claim for ancient life on Mars comes from a single meteorite, "ALH84001," named for where and when it was found—Allan Hills ice field, Antarctica, in 1984—together with its official sample number (001). The 1.9 kg softball-sized rock was plucked out of the ice by Roberta Score, a graduate of my department at UCLA and member of an annual expedition of the US National Science Foundation's Antarctic Search for Meteorites Program.

Dates measured by **radioactive** minerals show the rock formed nearly the same time Mars was born, 4.5 billion years ago, probably a few kilometers deep within the congealing crust. Though sketchy, its subsequent history can be pieced together.

Like the early Earth, ancient Mars was bombarded by rocky chunks swept from orbit as it circled the Sun. By one scenario these impacts cracked and fractured ALH84001, and since this was early in Mars' history, 3.6 billion years ago, when the planet was warmer and wetter, groundwater swept through the fissures and filled them with carbonate mineral. But the age of the carbonate fracture-fillings is open to question, by some evidence dated from only 1.3 rather than 3.6 billion years. And while no one disputes the veins are packed with carbonate, an alternate version has it emplaced much later when the impact that careened ALH84001 off the Martian surface infused the veins with hot CO_2-charged fluids.

About 16 million years ago an **asteroid**, a huge meteorite, struck Mars with terrific force, gouging a large crater and ejecting pieces of Mars' surface with enough power to escape its gravitational pull. ALH84001 was one of those pieces. It hurdled through space for millions of years until it felt Earth's tug and fell to

Antarctica 13,000 years ago.

Rocks like ALH84001 are rare as hens' teeth. Though thousands of meteorites are known to science it is one of only 12 meteorites identified as Martian. A chemical signature (a mix of the **isotopes** of oxygen, $^{16}O/^{17}O/^{18}O$) shows they are not Earth rocks and not from the Moon, and because they share the same chemistry all are thought to come from the same source. Like the others, ALH84001 is an **igneous**, once-molten rock, so it and the others must have formed on a body large enough to have partly melted. Planet-sized bodies fill the bill, and with the Earth and Moon ruled out only Mars, Venus, and Mercury are left.

The link chaining the group to Mars is provided by one of ALH84001's siblings (meteorite EETA79001) which contains tiny pockets filled with gases that match those measured in Mars' atmosphere by NASA's 1976 Viking landers. The gas mix is distinctly Marslike and differs from any known elsewhere.

Rocks Trickle in from Mars

Most meteorites are debris left over from when the solar system formed. But about two dozen have been identified as chunks dislodged from planetary neighbors, half from Mars, half from the Moon. Six of the 12 Mars meteorites were discovered in Antarctic ice fields, so of the 8,000 meteorites recovered from Antarctica roughly one of every 1,000 is a piece of Mars.

Though ordinary travel-times from Mars to Earth are millions of years, under some conditions they can be very much less. Using computers, scientists at Cornell University simulated the histories of more than 2,000 objects careened off Mars' surface and found that a small fraction could have arrived in no time flat. According to their calculations, "fast transfers (taking less than a year) from Mars to Earth must have occurred numerous times during the Earth's past....If Martian microorganisms can survive a year in space, many may have already arrived." The Cornell group was concerned with *live* organisms whereas the Mars Team's evidence is of life long dead. If live microbes could get here, their fossils might too!

SEARCH FOR THE SMOKING GUN

What Is Known and What Is Not?

The claim for ancient life on Mars is backed by three types of evidence, all found in the carbonate-filled fractures of ALH84001:

1. Pancake-shaped globules—orange-colored carbonate discs and their dark (iron sulfide and oxide) rims—made up of minerals that on Earth can be formed by bacteria.
2. Organic molecules like those produced by breakdown and geologic aging of fossilized organic matter.
3. Minute threadlike and jellybean-shaped bodies that resemble earthly fossil microbes.

At the NASA press conference Administrator Goldin proclaimed the findings "compelling." In one sense of the word they certainly are, for like clues in a good detective story they are captivating, even gripping. But the findings are far from "irresistible, overwhelming," as the term is also used, and of the three lines of evidence only one, the possible fossils, is a potential smoking gun.

Martian Minerals

Consider first the mineral evidence. On Earth, bacteria sometimes play a role in forming **carbonate, sulfide,** and **oxide** minerals like those in ALH84001. Yet the same minerals are common products of geology, made by wholly inorganic means, and are present in other meteorites where their nonbiologic source is beyond dispute. The minerals hold clues to the history of the Mars rock but are not firm evidence of life.

Nothing about the carbonate discs pegs them as products of biology. Resurrected from their earlier designation as possible "Martian protozoans," their link to bacteria is equally unproven. The sulfides also lack a biologic signature and the mix of sulfur isotopes they contain would on Earth tag them as inorganic rather than bacterial. And though the iron oxides (minute crystals of magnetite, Fe_3O_4) have been dubbed "**magnetofossils**" in the popular press and some scientific articles, they are fossils in name only. Crystals like those in ALH84001 are present in certain strains of microbes which use them as tiny compasses if they are the right size (40 to 120 nm), have their magnetic poles pointing in the same direction, and are linked in long ensheathed chains that boost the magnetic signal. The iron oxides in the Martian meteorite come in other sizes as well and are randomly oriented, never bound in chainlike aggregates, and cannot be told apart from particles formed without life.

A piece of the puzzle not yet in place is the temperature at which the minerals formed, crucial because life's chemistry breaks down if conditions are too hot. The current World's Record is 113° C held by superheat-loving microbes ("**hyperthermophiles**") isolated from deep-sea fumaroles, and key molecules of life begin to fall apart at temperatures above about 125° C. The Mars Team argues that the chemistry of the vein-filling carbonate shows it formed at temperatures low enough (less than 80° C) for life to exist. But other workers using different indicators arrive at much higher estimates—150°, 250°, 300°, even more than 650° C—temperatures too hot for life. A high temperature would rule life out whereas a low one would show only that microbes could have existed, not prove they did.

New results can be expected from continuing studies of the mineralogy of ALH84001 and its Martian siblings. But unless the minerals are somehow shown definitely biological they cannot answer the question of life on Mars.

Organic Molecules Possibly from Mars

One of the most exciting findings of the Mars Team is the identification in ALH84001 of organic compounds known as **polycyclic aromatic hydrocarbons**, in shorthand, "PAHs." Though this find points in a promising direction for future exploration of the Red Planet, the presence of PAHs, like that of the minerals, is not proof of Martian life.

It seems odd to think that chemicals termed "organic" could be other than signposts of life. But "**organic**" and "**biological**" do not have the same meaning. Organic compounds are composed of carbon and hydrogen (combined often with oxygen and/or nitrogen) and can be made by either biological or nonbiological processes. The confusion comes because the same four elements, CHON, make up biological molecules, biochemicals formed by the biochemistry of living systems.

All biochemicals are organic but not all organic chemicals are biological. Life itself emerged from a primordial soup of organic but entirely nonbiological substances. Organic (but nonbiological) CHON-compounds are actually quite common throughout the Universe, made by inorganic chemical reactions powered by starlight. PAHs are organic because they are made of carbon and hydrogen. But because they are not made in living cells they are not biochemicals, not molecules formed by life.

Organic molecules buried and pressure-cooked in rocks undergo a series of chemical changes that lead slowly toward ring-shaped molecules of pure carbon (graphite, or if the pressure is especially intense, diamond). PAHs, ring compounds made of carbon and small amounts of hydrogen, are a way-station along this path. The same path is traveled by all organic compounds regardless of their starting makeup, regardless of their (biological or nonbiological) source. So PAHs are abundant in the ancient organic matter of fossil plants and animals, but are common also in automobile exhaust and factory smoke (generated from burning of fossil fuels) and present even in the vapors rising from a grilled steak. Though PAHs are not biochemicals, not formed by life, on Earth they come from breakdown of once-living matter.

Yet PAHs are made easily by simple chemical reactions in the total *absence* of life so they are common also in meteorites, especially those rich in nonbiological organic matter, and abundant on surfaces of dust grains and tiny graphite particles that float through interstellar space. Like PAHs in other meteorites, those in ALH84001 may be entirely nonbiologic.

There is yet another problem. Ever since the NASA press conference there has been debate whether the PAHs found in ALH84001 actually belong to the Martian rock. There are good reasons to believe they were in the rock when it crashed to Earth, but it's not easy to be sure because PAHs from atmospheric pollution and probably also Antarctic coals are present in the snow and ice at Allan Hills and those in the meteorite are in cracks where they may have been deposited from seeping meltwater. However the debate turns out it cannot answer the question of past life on Mars. Because PAHs can come either from geologic breakdown of life (as on Earth) or from totally nonbiologic sources (as in meteorites) there is no way to tell whether the PAHs in ALH84001 are or are not signposts of life.

Still, if truly from Mars the PAHs could pave the way to important findings. NASA plans to hunt for life in Mars samples to be returned to Earth in 2008. Only hard-line enthusiasts think the dust and stones will harbor anything alive—unlike its more clement past Mars now is a terrible place to live, its surface drenched in lethal UV rays and so frigid there's no water for life to use. By one wag's account, "the odds of finding anything alive are slim and none, and Slim just left town!" So NASA has pinned its hopes on unearthing (unmarsing?) microscopic fossils, a needle-in-a-haystack hunt. The first task is to find the haystack and organic compounds like PAHs may provide the means. On Earth, organic matter and life go hand-in-hand. The same should hold for Mars if life ever existed there. NASA's best bet is to ferret out and bring back rocks rich in coaly carbon and search them for tiny fossils.

Fossil Microbes from Mars?

Ancient life on Mars. The minerals can't prove it. The PAHs can't either. The "fossils" could—but they don't, and there are good reasons to question whether they are in any way related to life.

"A picture is worth a thousand words." Shown in newspapers, magazines, and on TV around the world, "Mars fossils" have captured the public fancy. Their lifelike shapes are palpable, far easier to understand than arcane chemistry. But unlike the press, the Mars Team has handled the supposed fossils with kid gloves. Their seven-page article includes only four sentences on the objects which suggest blandly that as "features resembling terrestrial microorganisms...or microfossils [they are] compatible with the existence of past life on Mars." The objects they picture are exceedingly tiny, 20 to 30 nm across (Figure 5.8), and though their jellybean-like form resembles some bacteria they are much too small for the comparison to hold. More than a million times smaller than a run-of-the-mill bacterial cell (Figure 5.9) they are actually more like ribosomes, 20-nm-size protein-making bodies present in cells in prodigious numbers—about 20,000 in a typical bacterium, more than a hundred thousand in a human cell. The "Mars fossils" are the size of minute particles *within* bacteria, not bacteria themselves!

Other than shape, size is the only hard fact yet revealed about the fossil-like bodies, potentially telling because there is a limit to how small cells can be. The tiniest creatures on Earth, bacteria of the genus *Mycoplasma* that live as parasites in cells of other organisms, usually mammals, show the limits of life. The most minute are about 0.1μm (100 nm), contain only a fraction of the genes of a typical bacterium, and are encapsulated by a thin membrane rather than a sturdy cell wall (Figure 5.10). Truly tiny, about one-billionth as massive as a single protozoan, **mycoplasmas** are able to function because they are bathed in the nutrient-rich juices of the host cells they inhabit.

If the Mars "fossils" actually were once alive they must have been composed of cell-like compartments that cordoned off their living chemistry from the surroundings. But because their 20 to 30 nm-breadth is thinner even than the simplest bacterial cell wall they would have been like mycoplasmas, bounded by a membranous structure rather than a thick-walled casing. Judging from biological membranes which take up 6 to 10 nm, the living cavity would have been minuscule.

No information has been given about what encases the "fossils" (or even whether they have true cells) so the space available to house living processes can only be estimated. But it is certainly much less than that of the tiniest mycoplasma, evidently by about 2,000 times (Figure 5.10). In other words, for the "Mars microbes" to grow and reproduce even like rudimentary mycoplasmas each of their

SMALLEST (20 to 30 nm-diameter) MARS "MICROBES"

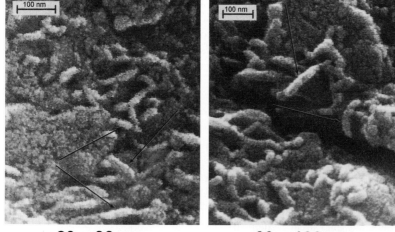

20 x 90 nm   **30 x 130 nm**

FIGURE 5.8   Jellybean-shaped objects (at arrows) pictured in the Mars Team's article of August 1996.

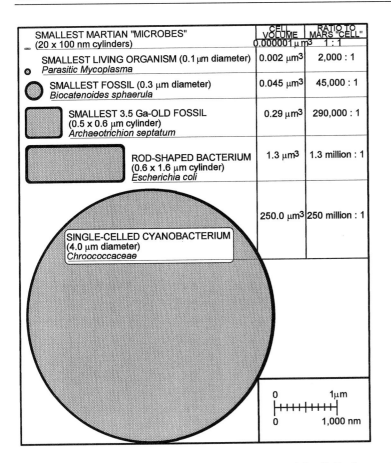

	CELL VOLUME	RATIO TO MARS "CELL"
SMALLEST MARTIAN "MICROBES" (20 x 100 nm cylinders)	0.000001 μm³	1 : 1
SMALLEST LIVING ORGANISM (0.1 μm diameter) *Parasitic Mycoplasma*	0.002 μm³	2,000 : 1
SMALLEST FOSSIL (0.3 μm diameter) *Biocatenoides sphaerula*	0.045 μm³	45,000 : 1
SMALLEST 3.5 Ga-OLD FOSSIL (0.5 x 0.6 μm cylinder) *Archaeotrichion septatum*	0.29 μm³	290,000 : 1
ROD-SHAPED BACTERIUM (0.6 x 1.6 μm cylinder) *Escherichia coli*	1.3 μm³	1.3 million : 1
SINGLE-CELLED CYANOBACTERIUM (4.0 μm diameter) *Chroococcaceae*	250.0 μm³	250 million : 1

FIGURE 5.9   Comparison of the sizes of living and fossil Earth organisms and the "microbes" reported in the Mars Team article.

biochemicals would have to do the work of some 2,000 earthly counterpart molecules.

Something is amiss. How could such tiny microbes live? They're said to be rock dwellers, not parasites, not bathed in life-supporting nutrients. Claimed to be "primitive," it's difficult to fathom how their biochemistry could be so efficient (especially in comparison with a mycoplasma's which is actually a highly evolved cut down version of that in a larger originally free-living ancestor). And if they are billions of years old and too tiny to have cell walls why are aren't they crushed, flattened, torn to bits like ancient fossils on Earth?

At the NASA news conference pictures were unveiled of other "fossils," more photogenic and somewhat larger than the ones mentioned in the article. Two types were shown, curved simple cylinders (Figure 5.11) and a single stringlike specimen cracked into segments (Figure 5.12). Though similar in general shape to earthly fossils their volumes are hundreds of thousands of times smaller.

Hardly any scientific data are available about either these or the jellybean-shaped structures. There is no evidence that shows what they are made of. Or whether they are solid, or hollow, or composed of cells. Or how numerous they are and whether they are present elsewhere in the rock. Or why they are so pristine, so perfectly preserved. Or why they lie on a fracture face rather than being embedded in rock as Earth fossils are. It hasn't even been shown they are actually part of the meteorite rather than earthly contaminants or ropy substances splashed on the rock

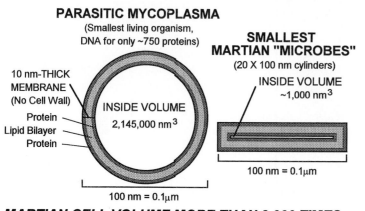

MARTIAN CELL VOLUME MORE THAN 2,000 TIMES SMALLER THAN SMALLEST LIVING ORGANISM

FIGURE 5.10   Comparison of the smallest organism on Earth (parasitic mycoplasma) with the "microbes" reported in the Mars Team article.

fragment when it was examined.

The shape of the objects suggests they may be biological while their size argues they almost certainly are not, but there are so many unanswered questions the issue cannot be decided. Within months after their article was published the Mars Team came to the same conclusion, in their words that "the morphology [shape and size] of the possible fossil forms ... is certainly not definitive, and more data are needed."

Could Bizarre Mars Fossils Be Identified?

Because bacteria are the most ancient forms of life on Earth and can live almost anywhere it seems sensible to search Mars rocks for bacteriumlike fossils and signs of their living processes. But what if Martian "bacteria" differed markedly from those on Earth? Could fossil cells truly bizarre in earthly terms be identified as remains of life?

This type of question is not new to paleobiologists, especially those hunting life's remnants in ancient (**Precambrian**) rocks. The answer is yes, even for tiny organisms long extinct that bear no obvious relation to life today.

Two examples illustrate the point. Imagine a minute microbe having the form of a soccer ball covered by a layer of golf balls encapsulated by a basketball. Nothing so bizarre exists today. But cells like this floated in shallow seas 2,100 million years ago (Figure 5.13). Named *Eosphaera* ("dawn sphere") and known to be planktonic because of its spread in the fossil-bearing rock (the Gunflint Formation of southern Canada), we can only guess that the soccer ball core is a central cell, the golf balls reproductive bodies, the basketball a protective shroud. But we know for certain that *Eosphaera* was once alive—it's made of organic matter (now coaly), has cells and wall layers, is known from many specimens (some complete, others decayed, distorted, torn, flattened), has a biological size-range, is part of a complex biological community, was fossilized by processes well understood.

A form even more other-worldly is shown in Figure 5.14—a chocolate-

CYLINDRICAL FILAMENTS

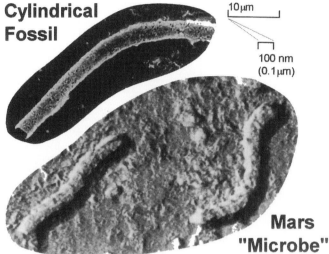

Cylindrical Fossil

10 µm

100 nm (0.1 µm)

Mars "Microbe"

FIGURE 5.11 A *bona fide* fossil (*Eomycetopsis robusta* from the 850 million year-old Bitter Springs Formation of central Australia; scale = 10 µm) and cylindrical Mars "microbes" (scale = 100 nm = 0.1 µm).

SEGMENTED FILAMENTS

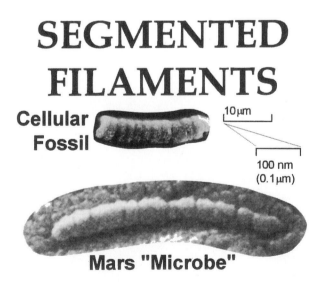

Cellular Fossil

10 µm

100 nm (0.1 µm)

Mars "Microbe"

FIGURE 5.12 A cellular fossil cyanobacterium (from the 850 million year-old Bitter Springs Formation of central Australia; scale = 10 µm) and the single known "segmented" (cracked) Martian "microbe" (scale = 100 nm = 0.1 µm).

covered peanut connected by a slender stem to a little umbrella. Bizarre indeed! But organisms of this form, too, have been found in the Gunflint rocks (Figure 5.15) and named *Kakabekia umbellata* (for the Kakabeka Waterfall, where the first of its type were found, and its umbrellalike crown). How *Kakabekia* fits in the Tree of Life is completely unknown, but enough specimens have been found to guess its life cycle (the crown expands from parasol to large umbrella as peanutlike bodies are spawned to reproduce the stock).

Life varies over time, place to place, no doubt planet to planet. But if the right questions are asked and enough data amassed, even fossils strange to us can be identified as remnants of life long past.

LESSONS FROM THE HUNT

Headlines Win

Perhaps the foremost lesson learned from this latest chapter in the search for life on Mars is one we've known all along: At least initially, *headlines and sound bites win while facts and reason lose.* Most Americans (more than 60% by one poll) agree that "NASA has proved primitive life was present on Mars."

In the face of iffy evidence and unanswered questions, why do so many take this view? Some simply want to believe, others are impressed by NASA's track-record and think its backing foolproof. Few are familiar with the researchers' facts and even fewer have seen their report. Yet hopes certainly have been stirred—sold as "good science," if the claim collapses it will give a black eye not only to NASA but science too.

The Humanness of Scientists

A second lesson is that *scientists are no more immune from workplace pressures than anyone else*, illustrated by two different readings of the recent history.

By one, geologist-mineralogists at the Johnson Spacecraft Center chanced on distinctive mineral structures they thought possibly biologic in an ancient Martian meteorite. Spurred by these hints they researched meticulously for two-and-a-half

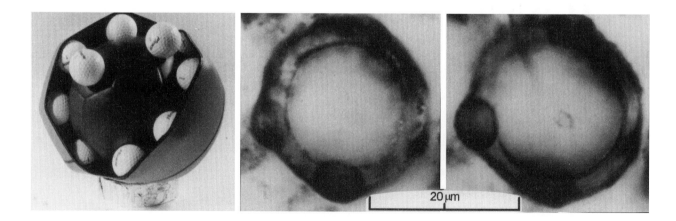

FIGURE 5.13 A model (left) of the bizarre 2,100 million year-old microfossil *Eosphaera tyleri* (center and right).

FIIGURE 5.14   A model of the 2,100 million year-old fossils shown in Figure 5.15.

years and added supporting evidence from the Stanford PAHs-group and Georgia and McGill University specialists. Judicious scientists, they released the findings only after their soon-to-be-published manuscript passed **peer review** (a scrutiny by fellow scientists for accuracy and completeness) and then only at the behest of NASA Headquarters which felt duty-bound to inform the public. The article was meant as a preliminary report, not the final word, and the claim was of evidence "compatible" with past life on Mars, not that they had proved it present. But "compatible…possible…perhaps…maybe" make mushy sound bites and don't sell newspapers. The Mars Team was done in by an over-zealous press corps.

An alternate scenario has it that the scientists came to be so caught up in the find that normal caution was cast aside. Evidence at odds with the hoped-for outcome was marginalized (such as that of life-searing high temperature and the unbelievably small size of the putative fossils) or even shoehorned to fit the story (such as the "protozoans" recast as bacterial detritus). To seal the case, eye-catching pictures of fossil-like objects were unveiled to the public without peer review or the backing of solid studies. The announcement of still-preliminary results was premature, but with congressional budget hearings in the offing the scientists acquiesced to higher-ups who wanted NASA in the headlines. President Clinton's introductory remarks and the press conference itself were part of an elaborate PR blitz. The event happened a week earlier than planned not so much because the news had leaked (as it surely had, at least from White House consultant Dick Morris to his ladylove at the Jefferson Hotel) but to avoid being upstaged by presidential candidate Bob Dole's impending announcement of a running mate.

Parts of each version may be right.

Science Is Not a Guessing Game

But there is a saving grace to this episode, embodied in a third lesson—*science is self-correcting*. There are fine lines between what is known, guessed, and hoped-for, and because science is done by real people these lines are sometimes crossed.

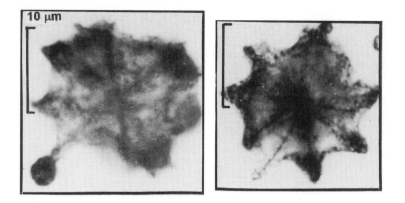

Kakabekia umbellata
GUNFLINT FM., CANADA, ~2,100 Ma

FIGURE 5.15 Specimens of a bizarre Precambrian fossil (*Kakabekia umbellata*) not obviously related to microbes living today.

Yet even if science's "hard rules" are more slushy than generally assumed, it can and will get the answer right.

Science is not a guessing game. The goal is to know. Feel-good solutions do not count. Scheuchzer's *Homo diluvii testis* was either "Man, a witness of the Deluge" or it was not. (It wasn't.) Beringer's famous "fossils" were either works of nature or they were not. (They weren't.) As for Mars, life either once existed there or it didn't—and ALH84001 either contains telling evidence or it doesn't. At the end of the day, hard facts will sort it out.

It is right to demand that extraordinary claims be backed by extraordinary evidence. Yet in the hunt for life on other planets another Sagan catchphrase applies as well: "Absence of evidence is not evidence of absence." It there was once or is now life on Mars or elsewhere in the accessible reaches of space, science must find means to ferret it out!

Acknowledgments

I thank Jane Shen-Miller for helpful suggestions in the preparation of this essay which is adapted from *Cradle of Life* (J. W. Schopf, Princeton University Press) to be published in 1999.

Further Reading

Beringer, J.B.A 1726. *Lithographiae Wirceburgensis* (University of Würzburg, Germany: Mark Anthony Engmann), 130 pp.

Gladman, B.J. and Burns, J.A. 1996. Mars meteorite transfer: Simulation. *Science* **274**: 162.

Goldsmith, D. 1997. *The Hunt For Life On Mars* (New York: Penguin Books), 267 pp.

Jahn, M.E. and Woolf, D.J. 1963. *The Lying Stones of Dr. Beringer* (Berkeley, CA: University of California Press), 221 pp.

McKay, D.S., Gibson, E.K., Jr., Thomas-Keprta, K.L., Vali, H., Romanek, C.S., Clemett, S.J., Chiller, X.D.F., Maechling, C.R. and Zare, R.N. 1996. Search for past life on Mars: Possible relic biogenic activity in Martian meteorite ALH84001. *Science* **273**: 924-930.

McKay, D.S., Thomas-Keprta, K.L., Romanek, C.S., Gibson, E.K. and Vali, H. 1996. Search for past life on Mars: Technical comment, *Science* **274**: 2124.

van Regteren Altena, C.O. and Möckel, J.R. 1967. *Minerals and Fossils in the Teyler Museum* (Haarlem, The Netherlands: Teyler Museum), 48 pp.

CHAPTER 6

ARE WE ALONE IN THE COSMOS?

TOBIAS C. OWEN[1]

THE QUEST FOR A COSMIC CONNECTION

The desire to understand our own destinies in terms of some great cosmic design is one of the deepest human tendencies. And the "design" is much more engaging, more fascinating, if it includes superior beings who have an interest in us strong enough to open the possibility of communication if not actual visits. Because this feeling of cosmic loneliness is so human and pervasive, those of us interested in obtaining a scientifically verifiable answer to the question posed in the title of this essay find ourselves faced by a large body of received wisdom, purportedly authoritative words from sources that range from The Holy Bible to "they say." In our quest we must tread a narrow path, overgrown with many varieties of religious beliefs, legends, myths, and speculation, only to find ourselves finally confronted by a large sign emblazoned with letters that spell out the unmistakable message "no admittance without hard evidence."

We live in an age hungry for myths. In ancient times, these wonderful stories of heroes, heroines, gods, and monsters provided a structure our ancestors could use to make sense of the world around them. For many people this essential function of mythology has been satisfactorily supplanted by science, a markedly different way of looking at the world that achieves a far better result. Despite the extraordinary successes of science, however, marvelous mythical creatures are with us still—in response to modern requirements, changed in form and habitat into visitors from space.

We can easily trace this transformation. In Homer's time, the Gods lived comfortably on Mount Olympus (Figure 6.1) while fierce monsters roamed about in remote parts of Greece and more distant regions of the Mediterranean world. One thinks of Scylla and Charybdis, for example, guarding the Strait of Messina between Sicily and Italy. Four hundred years later, Periclean Athens was flourishing and there were Greek colonies in Sicily. The old mythical monsters had been replaced by new ones in northern Europe and Africa, little known places that were far away. This retreat has continued, leaving us with tales of Yeti in the high Himalayas, Bigfoot in the dense forests of the Pacific Northwest, and of course

[1]Institute for Astronomy, University of Hawaii, Honolulu, HI 96822.

FIGURE 6.1 Beautiful Mount Olympus in northern Greece, legendary home of Zeus, Athena, and all the other deities of Greek mythology. A climb to the top is a thrilling experience. (Courtesy of K. Zolotas, Litochoron.)

Nessie, that enigmatic yet lovable reptilian that undulates about in the peat-filled waters of Loch Ness.

But the retreat has gone even farther as the human mind has cleverly located the ultimate haven for its fantasies. During the present century, our mythical companions have left the Earth completely and are now imagined as populating planets in our solar system and beyond where they exercise powers beyond human comprehension yet never forget their enduring interest in planet Earth and its floundering inhabitants.

In its purest form, this modern manifestation of mythology involves UFOs, malevolent Martians, and a massive reinterpretation of history that views an incredible array of signposts from the past—from the pyramids of Egypt to cave paintings by early humans—as strongly influenced by or even the work of visitors from the sky. Some authors have gone so far as to suggest that venerated classical myths are nothing more than the simple-minded responses of our ignorant ancestors to the apparently magical visits of galactic cosmonauts. The new mythology thus incorporates the old, in much the same way that the invading peoples who in time came to be the classical Greeks absorbed into their own pantheon some of the legends and deities of their conquered foes.

In this chapter I will briefly review the UFO phenomenon and the myth of ancient astronauts, and then focus on Mars and our current scientific understanding of why that planet is so different from Earth. Is the Earth the only inhabited planet in the Universe? Many scientists don't think so and we will see why, using the famous "Drake Equation" to organize our ideas. If we are ever to find a scientific answer to whether we are alone in the Cosmos, we need do some experiments—and the good news is we have already started!

THE MYTH OF UFOS

If we are *not* alone in the Cosmos, it might seem reasonable to expect aliens from one or another of our nearest neighbors to pay us a visit. This seemingly reasonable expectation has gained such force that the opposite argument has also been advanced, the notion that the absence of such visits would tell us that we are in fact alone.

Aside from the simple observation that "absence of evidence is not evidence of absence," as Carl Sagan liked to remind us, it's worth considering what evidence, if any, there actually is. In fact, throughout recorded history there *have* been reports of visitors coming from the sky. These are easily understood since the heavens are frequently viewed as the natural home of deities who by definition are interested in us and would be reasonably expected to visit or send their messengers to us from time to time. In Biblical time, wings or a chariot of fire were the preferred means of conveyance, but today it is a spaceship. Tubular or saucer-shaped, these mechanical marvels are said to swoop through our skies flashing their lights and occasionally landing to take a closer look. Sometimes they are reported even to abduct humans, taking them for rides or conducting bizarre experiments on them. If these tales seem a bit beyond the pale you can read about them yourself—they pepper the Internet and show up almost weekly in the tabloids at your local supermarket!

Is there any truth to such stories? Undeniably, people *do* see objects moving across the sky they cannot identify, an experience reported by astronomers and physicists, clergymen, police, and airplane pilots, even a president of the United States. The sightings of UFOs (*Unidentified* **F**lying **O**bjects) are genuine, but what is it those who report them actually see?

It usually turns out that perfectly natural objects are being mistaken for something mysterious. Planet Venus heads the list. At its brightest Venus can cast a shadow on the Earth! It is exceptionally luminous and to people unfamiliar with the sky its appearance can seem quite extraordinary (Figure 6.2). Seen through broken clouds or wind-stirred tree branches, Venus can appear to be moving, rapidly and erratically like a high-powered spacecraft. Police have chased it along country roads, jet pilots have flown after it, the navy has even fired guns at it. Fortunately, Venus is far enough away to have survived these indignities!

Birds flying at night are also frequently "unidentified" as are artificial satellites, meteors, and clouds (Figure 6.3). You can be assured that almost any UFO having flashing red and green lights is an airplane, yet reports of many such sightings have been filed.

Amidst all the natural explanations there are problems of frauds and jokes. Pictures of flying "Frisbees" and garbage can lids, out-of-focus light bulbs, reflections of interior lights on windows looking out to the sky—all have been presented as evidence of alien spaceships. And mysterious "crop circles" cut into farmers' fields and celebrated as substantiating visits of galactic visitors have turned out to be the work of pranksters.

The sad truth is that we have no solid evidence for visits by spaceships. We have pictures of meteors carving their way through the Earth's upper atmosphere

FIGURE 6.2 The planet Venus is shown here, with the crescent Moon below and to the right, in the dawn sky above an astronomical observatory on Mauna Kea, Hawaii. This bright planet has been the source of many UFO reports. (Courtesy of David Morrison.)

and records of airbursts made by impacting interplanetary projectiles, fragments of rock and ice left over from the formation of the solar system. We have photographs of houses and cars damaged by meteorites and the actual stones that did the damage, now in our museums along with rocks that reached the Earth from the Moon and even from Mars. We have debris from our own rockets and satellites that have crashed back to Earth. *But we do not have a single fragment or picture of a spacecraft from another world.*

All we have are reports. Many are interesting, and some come from entirely reputable sources, but not one is supported by hard evidence. The backing is not even close to the kind that can convict a murderer and far less even than one looks for when buying a used car! If we truly want to know the answer, we need verified proof before we can accept the wonderful idea that the Earth is being visited.

What about a conspiracy? Devout believers in alien visits often suggest that the government has the "real evidence" which it won't release. It is certainly true that the U.S. government keeps secrets (although many of them finally come out). For UFOs, however, we are talking about a worldwide, multigovernment conspiracy. There is no reason to expect that alien spaceships land *only* in the U.S. If these spacecraft came all the way to Earth from some distant star and there are so many of them, they surely would have visited many countries. To make the conspiracy story stick we would have to assume that the governments of all countries have somehow banded together to deny their citizens this knowledge. Imagine India, Iran, Israel, Iraq, Indonesia, Ireland, Italy, and Iceland, all locked in a conspiracy to prevent us from knowing about extraterrestrial visits. I think we can agree that any such conspiracy would be most unlikely!

It would be wrong, however, to view the apparent lack of contact with extraterrestrials as discouraging or unexpected. The distances between even "close-packed" stars are immense, so it is not so surprising we do not receive daily visitations, even if the galaxy is teeming with life. To put matters in perspective, our fastest spacecraft, Voyagers 1 and 2, would take 100,000 years to reach the nearest star (even if they were headed directly for it, which they are not). To hop from star to star in times short by human standards one would need to move at **relativistic speeds** (that is, nearly as fast as the speed of light) which would take huge amounts of energy. In a beautiful example worked out by Nobelist Edward Purcell, a relativistic trip to a star only twelve light years away would require annihilation of a mass of matter and antimatter equivalent to the Empire State Building! And Purcell scrupulously ignores the engineering difficulties of the feat, such as the energy required to make that much antimatter and the need to isolate it from ordinary matter until the two meet in the unimaginable engine.

When one considers that messages of any length can be transmitted across our entire galaxy at the speed of light and a cost of many orders of magnitude less energy—something even our primitive civilization can do right now—it seems reasonable to think that even highly advanced civilizations would choose to communicate by messages rather than direct travel. But, of course, we don't actually *know* that! All of us must keep open the possibility there are some deep secrets of nature yet uncovered that might make space travel easy. And in the absence of such discoveries we can do experiments. We can test the idea that the Earth has repeatedly been visited by extraterrestrials who left signs of their visitations behind.

THE MYTH OF ANCIENT ASTRONAUTS

If we have no current visitors from the Great Beyond, what about visits in the past? As a first step, we can search ancient human records to see if we discover something that is out of place—some indication of knowledge that contemporary civilizations could not possess or some artifact or accomplishment that must have required outside intervention.

The greatest champion of this approach is Erich von Däniken whose book *Chariots Of The Gods?* sold millions of copies around the world in the 1970s and is still in print today. Von Däniken assumes that his readers have a very low level of scientific understanding—so low they will not notice his many inconsistencies and not check his extravagant claims. His is a perfect example of the initiating and spreading of a modern myth.

According to Von Däniken, there is a cave-drawing at Tassili in northern Africa that is an accurate representation of an astronaut in a space suit. He reproduces the drawing in his book and even credits the man who discovered it, Henri Lhote. The figure is an outline drawing of human without a neck (Figure 6.4A). A

FIGURE 6.3   A UFO abduction (or why astronomers work at night). (A) In mid-afternoon, a huge alien "spaceship" (a majestic stratus cloud) settles down on the volcanic summit of Mauna Kea, Hawaii, dwarfing the astronomical observatories there. (B) The "ship" flies off into the distance as screams of the abducted observatories' day-crew members grow steadily fainter. (Courtesy of Wayne Holland.)

quick trip to the library reveals that the drawing is one of several reproduced by Lohte's expedition and that other drawings show similar figures that are clearly barefooted, decorated with feathers, and holding bows (Figure 6.4B). All are simply rough representations of humans of the time. Depictions of local animals by the same early artists are also crudely drawn. Later inhabitants of the Tassili caves produced more accurate figures, some quite beautiful, and the later drawings include no figures resembling cosmonauts or their craft.

Von Däniken also writes about a cave-drawing discovered in California that to him looks like a depiction of a slide-rule. Since the primitive artists would never have manufactured slide rules of their own, he argues they must have seen one in the hands of visitors from an advanced civilization. One problem with this fantasy is that in the less than two decades since publication of Von Däniken's book, slide rules have disappeared, superseded completely even in our own rather simple

FIGURE 6.3 *Continued*

version of an advanced civilization by hand-held electronic calculators. Today, most young scientists and engineers have neither seen nor even heard of them!

A strange pattern of lines found on the dry and desolate plain of Nazca in Peru is claimed by by von Däniken to be a landing field for interstellar spaceships. Here he seems to assume that visitors from nearby stars bridge cosmic distances in fixed-wing aircraft—interstellar biplanes, perhaps?

And so it goes as von Däniken sometimes strays perilously close to the language of science. Is it really a coincidence, he asks, that the height of the great pyramid of Cheops multiplied by a thousand million is nearly equal to the distance from the Earth to the Sun? The ancient Egyptians wouldn't know this distance so someone must have told them, and perhaps that someone actually built the giant pyramids, a feat our author doubts the Egyptians could accomplish on their own.

The pyramids are a marvel, as anyone who has seen them can attest. They have inspired tall tales for millennia, including a fabulous account by Herodotus in the fifth century BC (when the great pyramids were already over 1,500 years old). Yet the quarries from which the blocks were taken and tools that were used are all still there. It is certain that these extraordinary royal tombs were built by the Egyptians and it is demeaning to them to suggest they needed help.

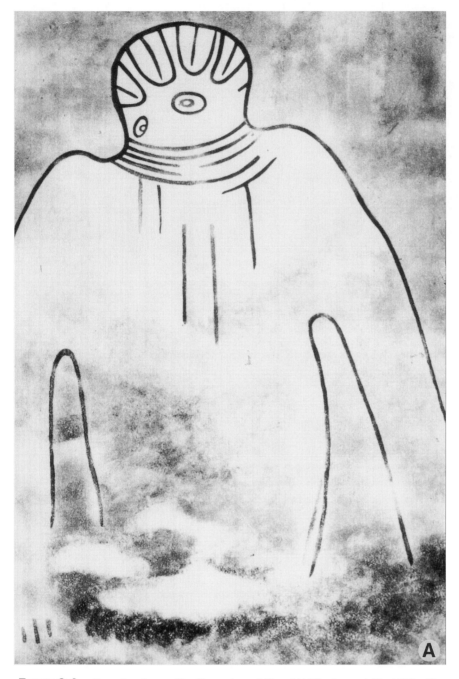

FIGURE 6.4 Cave-drawings at Tassili, northern Africa. (A) The figure dubbed "The Great God Mars" by its discoverer, Henri Lhote. Is this an alien in a spacesuit as Erich von Däniken suggests? (B) Other Tassali drawings from the same period as the "alien." Note the bows and feathers, and the decidedly humanoid hands and bare feet of the "astronauts." (Courtesy of Henri Lhote.)

But what about the height of the great pyramid and its claimed relation to the distance from the Earth to the Sun? Well, if ancient astronauts *were* responsible, why didn't they do a better job? Why isn't the height a more precise fraction of the Earth-Sun distance? After all, if you multiply the length of a common ballpoint pen (a "Bic Stick," for example) by a trillion, you get the distance from the Earth to the

FIGURE 6.4 *Continued*

Sun exactly! It appears that in von Däniken's claim we are indeed confronting a coincidence, and a not very remarkable one at that.

After many pages of this kind of "evidence," von Däniken winds up his case with an appeal to scientific authorities. Here he comes up with an interesting statement that bears consideration. He is discussing the equation developed by Frank Drake to estimate the number of intelligent civilizations in the galaxy (an equation we will discuss in more detail later). To impress readers of *Chariots Of The Gods?* with the seriousness of his claims, von Däniken writes (p. 141):

> Fantasy and wishful thinking may be concealed in all the deliberations and suppositions, but the [Drake Equation] is a mathematical formula and thus far removed from mere speculation.

Bona fide scientists know better. There is nothing magical about an equation, and von Däniken inadvertently proves this himself by suggesting that if one divides the height of the great pyramid into its area, one will obtain the dimensionless number π. (He should have used the pyramid's perimeter, not its area.) Equations are of course useful for showing relationships mathematically, but their terms have to be defined with care (as I will try to do when I later discuss the number of civilizations in the galaxy).

And if we're really serious about searching ancient records, why stop with recorded history? Why not search back through the rock record to a time nearly 4 billion years ago, a million times farther than van Däniken? We find wonderful evidence of former lifeforms and past events along this immense passage through time—fossil forests, ancient microbial menageries, diamonds made in the depths of the Earth, the concentrated iridium and carbon soot layer associated with the extinction of the dinosaurs—*but nowhere do we find fossilized remains of ancient spaceships or artifacts discarded by their crews.*

What can we conclude from all this? First, that people are so intensely interested in the modern myth of extraterrestrial visits they will accept almost any version of this exciting concept. Second, that we must never fail to apply critical thinking when confronted by an unusual idea. "How could I test that claim?" is a good question to keep asking yourself, whether someone is trying to sell you a used car, a political candidate, an explanation for the great pyramids, or a new scientific result.

THE MYTH OF MALEVOLENT MARTIANS

Of all the planets in our solar system, Mars is most like Earth. This was already apparent to astronomers over 150 years ago. They could not see the surface of Venus through their telescopes and the surface of the Moon never changed. But they saw that Mars has polar caps that wax and wane with the seasons, clouds that come and go—including magnificent globe-encircling dust storms—and dark regions on its surface, the intensities and outlines of which showed both seasonal and less regular changes. Here was a world like ours—smaller to be sure, and farther from the Sun, but blessed with an atmosphere and seasonal cycles that might even imply the growth and death of vegetation.

This attractive idea gained stunning support around the turn of the last century from the work of Percival Lowell, a wealthy amateur astronomer from Boston who established an observatory outside Flagstaff, Arizona, to pursue his interest in Mars. Lowell saw the planet crisscrossed with lines he interpreted as evidence of canals constructed by the planet's inhabitants to channel water from polar caps to desert regions (Figure 6.5A). Thus was born the concept of "Martians," aliens who by now are well-established members of modern mythology, known far more widely than Zeus or Athena.

Lowell's ideas were attacked at the time and with increasing vigor in years following. Other astronomers were unable to make out Lowell's "canals," and temperature measurements showed Mars was far colder than Lowell had imagined. Nevertheless, serious panic erupted in the United States on the evening of October 30, 1938, when Orson Wells broadcast his famous radio play based on H. G. Welles' *War of the Worlds.* The broadcast purported to be an on-the-scene account of an invasion of Earth by Martians who after landing in New Jersey were proceeding to conquer the United States with poison gas attacks. The day after this remarkable delusion the *New York Times* ran a front-page story about the "wave of mass hysteria" that "seized thousands of radio listeners throughout the nation."

Now sixty years later, after spacecraft have flown past, orbited, and even landed on Mars, Lowell's wonderful canals have been shown to be the product of wishful thinking (compare Figure 6.5A and 6.5B) and we still have no evidence for the existence of Martians. Not only are there no artificial structures on the planet, but NASA's 1976 Viking landers were unable to find any evidence of microbial life or even traces of organic compounds in the soil (which would have been detected in amounts less than one part in a billion, a sensitivity sufficient to find the proverbial needle in a haystack).

And yet.... From orbit the Viking spacecraft confirmed the presence of dry riverbeds and other evidence of water erosion. Today the Martian atmosphere is so thin that liquid water cannot exist on the planet's surface, but in the distant past Mars was wet. Evidently the planet had an early period in its history when the atmosphere was thicker and the surface warmer than it is today. The age of this epoch can be determined from counts of impact craters in the floors of the wider river channels which compared with the density of craters on the Moon (whose ages we know from rock samples returned to Earth) show that liquid water existed on Mars earlier than 3.5 billion years ago. Since then, the surface of the planet has remained similar to the way it is today—cold, dry, drenched by solar ultraviolet light, extremely hostile to life.

But in that earlier time, during the first billion years of the solar system's existence, conditions on Mars and Earth apparently were not too different. So if the two planets were endowed with the same starting materials it is conceivable that life could have arisen on Mars at the same time it did on Earth, some 4 billion years ago. Mars subsequently evolved along a very different path. If life *did* begin there, presumably as microbelike forms, in order to survive it would have been forced to adapt to a deep underground environment, as have some forms of life on Earth. Otherwise, we can be sure that Mars now would be totally lifeless.

THE MYTH OF THE GARDEN OF EDEN

Why are the histories of Earth and Mars so different? Why is Earth evidently the only one of the nine solar system planets that harbors life? Indeed, exactly what is it that makes a planet "Earthlike"? If we can answer these questions we will be well on our way to answering the question posed in the title of this chapter. And these answers, if we can phrase them as fundamental principles, will allow us to extrapolate from our own solar system to planets elsewhere in the Universe.

For our purposes, an Earthlike planet is one like ours that has a nearly circular orbit, on which both dry land and water are present, and whose mass is large enough to hold an atmosphere throughout its history yet small enough that hydrogen atoms can escape from the atmosphere into space. It is on worlds like this that life can originate and evolve to intelligent forms with which we might hope to communicate over interstellar distances.

In our own system of planets, we have learned that Venus is too close to the Sun and thus too hot to have liquid water on its surface, while Mars is too small to hold a thick atmosphere throughout its history.

To understand the problem posed by Venus' closeness to the Sun, let us imagine how the Earth would be affected were it to be moved to Venus' orbit. Immediately, the Earth's surface temperature would rise and the oceans begin to evaporate which would pump water vapor, a heat-holding "greenhouse gas," into the atmosphere and cause an increased **greenhouse effect.** In turn, the increased greenhouse would cause the surface temperature to rise further and produce more evaporation, more water vapor, and even higher temperature. In other words, a positive **feedback loop** would be established and the system would become a "runaway greenhouse." Eventually, the Earth's oceans would boil and all the water would be in the atmosphere which would be so hot that water vapor would rise to altitudes where its molecules would be broken apart by solar ultraviolet light.

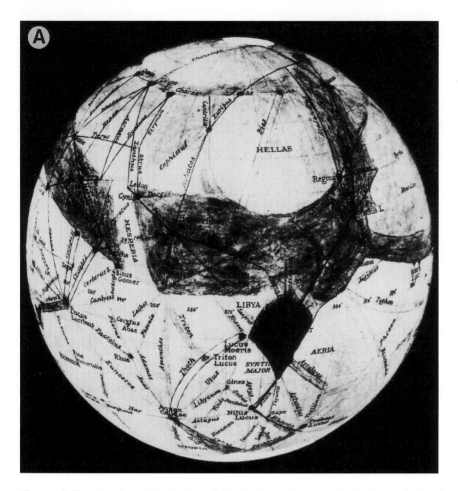

FIGURE 6.5   Canals on Mars? (A) A globe of Mars as drawn by Percival Lowell, showing the famous "canals" he thought he saw (the network of lines running from the polar regions toward the equator). (Courtesy of Lowell Observatory, Flagstaff, AZ.) (B) A mosaic of pictures taken by Viking in 1976 of the same hemisphere shown in Lowell's drawing. Though many of the major features depicted by Lowell stand out ("Hellas," "Libya," "Aeria," polar caps), his canals have vanished! (Courtesy of NASA.)

Hydrogen atoms from the water would then escape to space leaving oxygen behind which would combine with rocks. On Venus we know this actually happened because the water vapor remaining in its atmosphere is enriched in the heavy **isotope** of hydrogen by 150 times, compared to water on Earth, showing that a huge amount of hydrogen has escaped.

In the atmosphere of Mars we also have evidence from isotopes that a large fraction of the original atmosphere was lost. In this case it is the radioactive isotopes of inert, nonreactive, **noble gases** that give the clues: The excess $^{129}Xe/^{132}Xe$ and $^{40}Ar/^{36}Ar$ observed in the Martian atmosphere can be explained only by loss of original xenon and argon gases. The formation of carbonate minerals won't do it, and xenon is too heavy to be lost from the atmosphere by other mechanisms (such as thermal escape or sputtering). Loss by a massive process is required, and the intense meteorite bombardment that happened during the first 700 million years of the planet's history would do the trick.

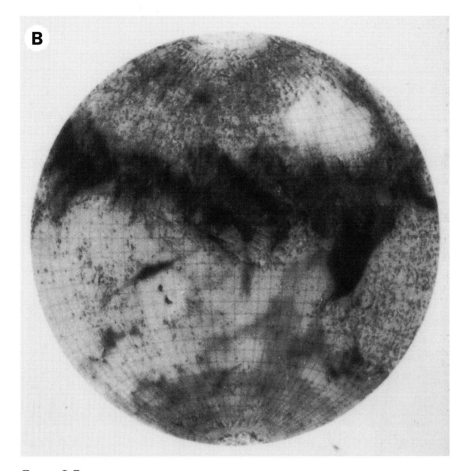

FIGURE 6.5 *Continued*

What does this tell us about the uniqueness of planet Earth? Is it possible that we are living in a cosmic Garden of Eden, that the Earth is so special we are the only sentient beings in the Universe?!

Obviously, the potential habitability of a planet is limited by its closeness to its star. No planet can be as close as Venus is to the Sun and have liquid water on its surface, an absolute essential for life. The outer limit for life—the farthest distance a life-bearing planet can exist from a star—is harder to define. If Mars had been the size of Earth or Venus it would have been massive enough to hold a thick atmosphere against the effects of meteorite impacts, as the other two planets obviously did, one with enough greenhouse effect to allow water to pool on the planet's surface, at least near the equator.

Where do atmospheric gases come from in the first place? Compared with Venus and Mars, Earth's present-day atmosphere is unique so perhaps the Earth was somehow special, blessed with gases that allowed the development of our unparalleled atmosphere, oceans, and life. Present evidence, however, suggests this is not so. The amounts of carbon and nitrogen in the atmosphere of Venus are very similar to Earth's and the abundance pattern of Earth's noble gases is close to that of Mars. As a result, most scientists think that on all three planets these gases were contributed by a flux of comets and meteorites early in their histories, events that left the scars of the bombardment we see still on the surfaces of the Moon, Mercury, Venus, Mars, and every other ancient solid surface in the solar system.

Comets can bring in water and all of the chemical elements on which life is based, and by doing so endow a planet with enough starting materials for life to gain a foothold. In our solar system the episode of heavy bombardment seems a natural part of planet formation, so we can expect the same sorts of starting conditions for life's origin to exist on Earthlike planets throughout the Universe.

Stunning confirmation of this idea has come recently through discovery of organic compounds in an ancient rock from Mars, a meteorite known as ALH84001. As discussed in Chapter 5, these compounds together with grains of iron oxides (the mineral magnetite) and iron sulfides, globules of carbonate, and tiny microfossil-like structures have been touted as possible evidence of ancient biological activity on the planet. Yet even if it is eventually shown this Martian meteorite does not contain firm evidence of life, the rock carries an important message. The organic materials (if not earthly contaminants) show that conditions for starting life actually existed on the only other planet in our solar system known to have had liquid water on its surface during the critical first billion years when life got started on Earth (Figure 6.6). This is strong support for the idea that given the right conditions, life may be fairly common in the Universe.

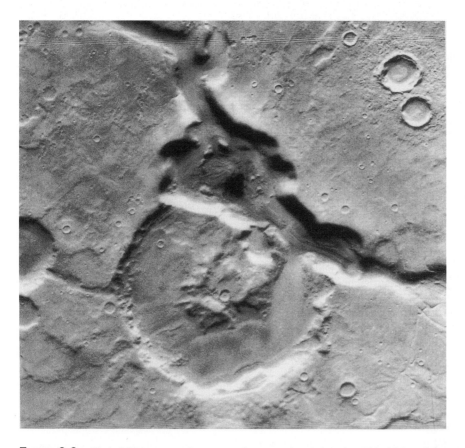

FIGURE 6.6   Early Mars was much wetter and warmer than it is today. This 35 km-diameter impact crater was breached by a water-carrying channel earlier than 3.5 billion years ago. Sedimentary rocks and debris deposited inside the crater from the running water can be seen where the wall was breached. The crater must have been a lake at the time water was flowing on Mars, a likely place for life to originate. (Courtesy of NASA.)

HOW MANY INTELLIGENT CIVILIZATIONS EXIST IN THE GALAXY?

We are now ready to try to muster a scientific answer to the question forming the title of this chapter, "Are we alone in the Cosmos?" For our guide we will rely on the **Drake Equation,** a formula developed by astronomer Frank Drake to estimate "N", the number of advanced civilizations in our galaxy at a given time:

$$N = R^* \cdot f_p \cdot n^e \cdot f_\ell \cdot f_i \cdot f_c \cdot L$$

To arrive at an estimate of "N," seven factors are multiplied together:

R^* = the rate of star formation in the galaxy;

f_p = the fraction of stars that have planets;

n^e = the number of Earthlike planets in such planetary systems;

f_ℓ = the fraction of such planets on which life develops;

f_i = the fraction of life-bearing planets that produce intelligent life;

f_c = the fraction of intelligent life-bearing planets that produce a civilization capable of and interested in interstellar communication;

L = the average lifetime of such civilizations, measured in years.

Despite the clarity and seeming power of the equation, it has only been possible to assign a precise numerical value to the first term of the equation, the rate of star formation! New stars form at a rate of about one per year, so "R^*" is approximately equal to unity. Numerical values for the next two terms, "f_p" and "n_e", are much less constrained. Thanks to the recent work of Michel Mayor and Didier Queloz in Europe and Geoff Marcy and Paul Butler in the United States, we are for the first time certain that extra-solar system planets actually exist (Table 6.1), but we are still not in a strong position to assign a number to "f_p", the fraction of stars that have planets, since the new technique applied by these workers has not yet been used to make a general survey. Nor do we have a firm basis to evaluate "n^e", the number of Earthlike bodies among those planets. The technique used so successfully by the European and American scientists is well-suited for finding giant planets but is simply not capable of detecting small ones having Earthlike masses.

So, to estimate the number of Earths present in the galaxy we are forced to resort to the indirect arguments given earlier. These arguments suggest that it is reasonable to assume one "Earth" for every three planetary systems. As we have seen, problems presented by the distance between a planet and its star are not so acute if the planet is of appropriate mass. Thus, if we assume that every planetary system has at least one inner planet with a mass similar to Earth's, since there are three in ours—Venus, Earth, and Mars—we can guess that every third system will have at least one planet of the "right size" in the "right place."

At present, however, we have no way to know whether there are actually *any* extra-solar system Earthlike planets where life could exist! One of the prime issues is whether liquid water—a necessity for life as we know it—exists on the surface of planets beyond our solar system. The popular press quickly picked up on the fact that among such planets recently discovered, some are located near enough to their stars that their surface temperatures would be higher than the freezing point of water (Table 6.1). And if liquid water could exist on these brave new worlds, why not life as well? The argument founders on where the liquid water would be. In our solar system, each of the giant planets (Jupiter, Saturn, Uranus, Neptune) has levels in its *atmosphere* where temperatures are high enough for droplets of liquid water to be present—since their atmospheres get thicker and hotter from the outside in it is only necessary to descend into the atmosphere far enough and one is guaranteed to find a level with tropical temperatures!

TABLE 6.1

Extra-solar system planets (known in March, 1997), with Mercury, Earth, and Jupiter for comparison.

Star	Planet's minimum mass (in units of Jupiter's mass)	Average planet–star distance (in units of the Earth–Sun distance)	Orbital period (in units of Earth days)	Orbital eccentricity
70 Virginis	6.6	0.43	116.6	0.40
55 Cancri #2	5	about 7	about 5,500	0.0
Tau Bootis	3.8	0.046	3.31	0.0
47 Ursae Majoris	2.4	2.1	1,090	0.03
16 Cygni B	1.7	2.2	810	0.57
55 Cancri #1	0.84	0.11	14.76	0.0
Upsilon Andromedae	0.68	0.04	4.6	0.0
51 Pegasi	0.47	0.05	4.23	0.0
Sun (and Mercury)	0.0002	0.39	116	0.21
Sun (and Earth)	0.003	1.0	365.24	0.02
Sun (and Jupiter)	1.0	5.2	4,332.6	0.05

But none of our giant planets possesses a solid surface, so liquid water could exist only in cloud layers like the cumulus clouds we're used to seeing in the skies of Earth. Because all the newly discovered planets (listed in Table 6.1) are giants, most more massive than Jupiter, we may safely assume that even those close to their stars will have liquid water only in clouds (Figure 6.7). Unless one is willing to entertain the idea that life on these planets exists solely as atmospheric "floaters"—the equivalent of Aristophanes' Cloud Cuckoo Land—the planets cannot be considered "habitable" like the Earth.

There is another problem. If the newly discovered giants that are so close to their stars came to their present locations by moving inward from more distant orbits around their suns they will have swept away any small truly Earthlike planets that were already in place—either gravitationally gobbling them up themselves or forcing them to merge into the central star.

But the final story about these newly discovered extra-solar system giants has yet to be told. New high-resolution spectroscopic observations of one circling the star known as 51 Pegasi suggest that it may be a peculiar oscillation of the star itself that causes us to interpret the signal as evidence of a planet very close to its star. This could explain away the problem of giant planets being "too close" to their stars but would not account for the observations indicating the presence of more distant planets. We need additional information, especially more high-resolution spectra, from these stars before we can reach a definite conclusion.

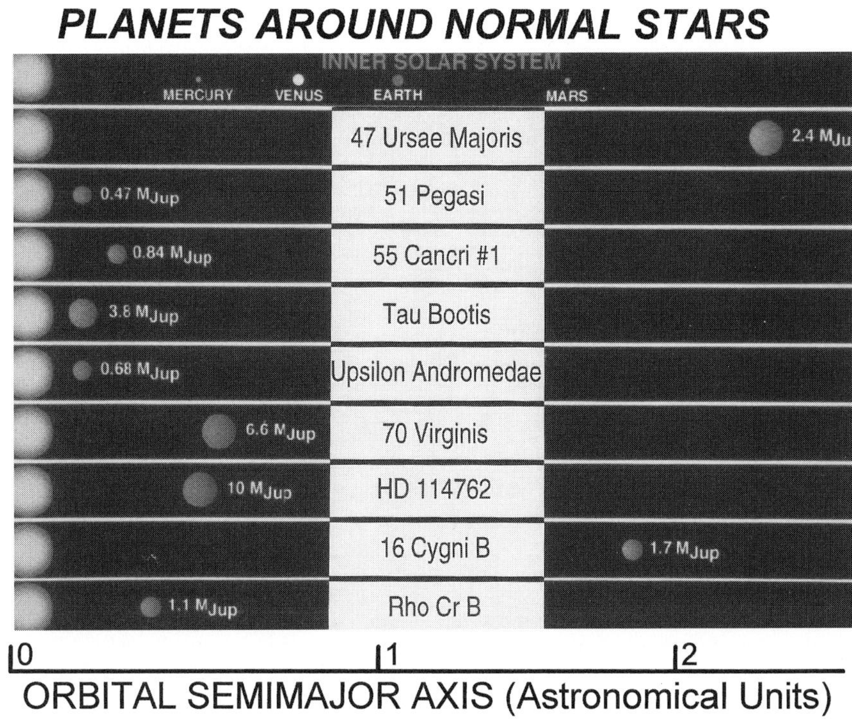

PLANETS AROUND NORMAL STARS

INNER SOLAR SYSTEM

MERCURY　VENUS　EARTH　MARS

	47 Ursae Majoris		$2.4\,M_{Jup}$
$0.47\,M_{Jup}$	51 Pegasi		
$0.84\,M_{Jup}$	55 Cancri #1		
$3.8\,M_{Jup}$	Tau Bootis		
$0.68\,M_{Jup}$	Upsilon Andromedae		
$6.6\,M_{Jup}$	70 Virginis		
$10\,M_{Jup}$	HD 114762		
	16 Cygni B		$1.7\,M_{Jup}$
$1.1\,M_{Jup}$	Rho Cr B		

0　1　2

ORBITAL SEMIMAJOR AXIS (Astronomical Units)

FIGURE 6.7　Several of the new planetary systems (see Table 6.1) are shown here in comparison with our own. (Modified after a figure provided courtesy of Geoff Marcy and Paul Butler.)

Meanwhile, we might ask about satellites as possible habitable worlds. Although none of the satellites of our own giant planets is massive enough to hold an atmosphere at Earth's distance from the Sun, perhaps the newly discovered giants have satellites with Earthlike masses encircling them. If so, especially those that orbit planets having masses greater than Jupiter (Table 6.1), would represent a set of potentially life-harboring bodies not ordinarily thought to have environments where life might arise and flourish. In the absence of additional scientific data, we will have to be satisfied to leave consideration of this part of the Drake Equation in this unfortunately speculative vein.

From the arguments we've just gone through it is reasonable to suggest that the product of stars that have planets times the number of Earthlike bodies in those planetary systems ("f_p" . "n^e") is probably less than 0.3 (that is, less than three in ten) and perhaps more than 0.03 (three in a hundred). Until we detect Earth-sized planets in other solar systems we simply won't know.

Moving to "f_ℓ", the fraction of Earthlike planets where life develops, we can see how important Martian meteorite AL84001 truly is. If we could prove that life once began on Mars we would have reason to feel confident that "f_ℓ" is approximately equal to unity. Without that knowledge, we just don't know—"f_ℓ" could be one, or one in a thousand, a million, even a billion!

The next term, "f_i" (life-bearing planets having intelligent life), is even tougher. Biologists are divided about the inevitability of the evolution of intelligence, once life gets going. Suppose there had been no great dinosaur-extinction event 65 million years ago? The dinosaurs had been around for 150 million years and (as far as we can tell) didn't seem to be getting much smarter. But after the demise of the dinosaurs, mammals *did* get smart. No one knows whether this was inevitable or just dumb luck.

We are perhaps on firmer ground in guessing that "f_c" (the fraction of intelligent life-bearing planets having civilizations capable of and interested in interstellar communication) is near unity. As we have seen earlier in this essay (and also in Chapter 4), curiosity about extraterrestrial beings is a remarkably deep-seated human attribute. Moreover, interstellar communication, even by the rather crude technology of radio, is not difficult. Once an intelligent species has emerged, it seems inevitable that it will attempt to communicate with hypothetical soul-mates elsewhere in the galaxy.

The estimates of "N" found by various authors who have gone through the exercise of multiplying these terms together vary over an enormous range—from "L" times one in ten billion ("L" . 10^{-10}) to "L" times one-tenth ("L"/10). It's easy to see why those who try to solve the Drake Equation look longingly at "L", the average lifetime of an advanced civilization, to compensate for possible low values among the other factors. And, in fact, "L" might be a very large number, even larger than the 150 million years achieved by dinosaurs. Or it might not. What lessons can we learn from ourselves?

Our own civilization first achieved ability to participate in interstellar communication about 50 years ago. This represents our value of "L", so far! We humans have obviously been around a lot longer than that, long enough to see some of the problems we must solve if we want to make "L" a truly large number. To the extent that we represent an average example of an advanced civilization—a notion dubbed the **Principle of Mediocrity** by I. S. Shklovskii and C. Sagan—we can assume that the same or similar set of "human problems" will be confronted by civilizations elsewhere in the galaxy.

In principle, the size of "L" is constrained by two general classes of difficulties—those caused by external agents and those we cause ourselves. The first category includes natural catastrophes such as a giant impact by an asteroid or a comet, a massive change in the Earth's climate brought about by some as yet unknown characteristic of the Sun, or the global outbreak of a deadly disease. Remarkably, we have achieved a level of scientific and technological maturity that allows us to cope with such events. We are steadily improving our ability to detect incoming asteroids and comets at distances far enough away to allow us to intercept and deflect or destroy them. The Sun may yet surprise us, but 4.5 billion years of experience—plus observations of many other solar-type stars and a lot of astrophysical theory—suggest that the Sun's luminosity will stay very constant for another 5 or so billion years. And despite past plagues and current concerns over AIDS, it appears that modern medicine is increasingly capable of coping with the danger of epidemics.

Although reassuring, these human capabilities carry with them the threat of leading to problems we cause ourselves. And while we have reached a given level of maturity we also have achieved the ability to destroy ourselves totally in a thermonuclear holocaust. Athough that danger seems to be diminishing, we are interacting with our environment in many self-destructive ways. The luminosity of the Sun may not change, but by pumping ever more carbon dioxide, methane, and fluorocarbons into the atmosphere we are increasing the greenhouse effect which inevitably will lead to changes in the global climate.

This, in turn, exposes perhaps the most serious threat we offer ourselves—a rampant increase in the human population (Figure 6.8). As Charles Darwin sagely noted: *"There is no exception to the rule that every organic being naturally*

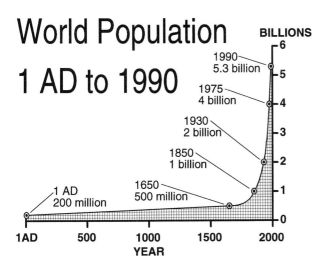

World Population 1 AD to 1990

BILLIONS

1990 5.3 billion

1975 4 billion

1930 2 billion

1850 1 billion

1650 500 million

1 AD 200 million

YEAR

FIGURE 6.8 The time it takes for the world's population to double has markedly decreased as the total number of humans has increased. Note that none of the great calamities of the 20th century—famines, epidemics, two world wars, Hitler's Holocaust, Stalin's purges—had much effect on the steady upward growth. (Data from *The World Almanac, 1991.*)

increases at so high a rate that, if not destroyed, the earth would soon be covered by the progeny of a single pair."

By steadily reducing threats to our existence, we have become the cancer that is destroying our planet, doubling our global population every 40 years! We *must* learn to regulate—and in actuality, to decrease—our population worldwide if we want our species to survive with a high level of technology for many more millennia. Otherwise we face the danger of being overwhelmed by social problems that could limit our own value of "L," perhaps to as little as 100 years.

BEATING MYTHS INTO SCIENCE

Let's end this essay by looking at the bright side of our new-found abilities. After centuries of speculation about extraterrestrial life, humans at long last have developed a technology that will allow us to find out whether or not advanced civilizations exist out there among the stars.

This is a new capability, a scant 50 years old, and in those five decades there have been only a few *deliberate* efforts to receive or transmit signals over cosmic distances. Television, FM radio, and radar transmissions, however, have all been spewing forth, heading out from our planet into the depths of interstellar space at the speed of light. Any inhabitants of planets around many of the bright stars we see in our skies have already been exposed to our signals. The number and sensitivity of our radio telescopes steadily increases, and new discoveries of telltale signals from atoms and molecules in our galaxy and its nearby neighbors continue to pour in. We have detected signals emanating from formaldehyde, ammonia, ethyl alcohol, and scores of other molecules in giant interstellar clouds. We have even discovered radio beacons beamed from stars smaller than the Earth, tiny remnants of supernova explosions that are composed only of neutrons, at least one of which has its own system of planetary bodies revealed by their gravitational effects on the star's rotation. However, in all this richness of received information there is not yet a shred of evidence of an artificial signal, something that would indicate the presence of an advanced civilization like our own.

Is it possible that here we are dealing with yet another myth? Perhaps "N" actually equals one, and we are it! This may turn out to be true, but we are in no position to reach a conclusion. Indeed, absence of evidence of extraterrestrial radio signals is as easy to understand as the absence of alien visitors. Sensitive as they are, our giant telescopes and their receivers can only search an exceedingly tiny fraction of the radio spectrum in a given direction in space at a given time. We do not know at what frequencies our hypothetical friends may be broadcasting, nor do we know from what direction(s) their broadcast(s) may be coming! Specialists in this field have overcome these problems by building receivers that can monitor literally tens of millions of frequencies simultaneously, feeding these signals through computers that can test them for evidence of alien intelligence. Such receivers coupled to dedicated radio telescopes would allow surveys of the entire sky as well as a search focused on the stars considered most likely to have planets like Earth.

Regrettably, intelligence on our own planet is often not as high as we might hope. Just a few years ago a NASA program to embark on a systematic scientific search for extraterrestrial signals, rightly called by Frank Drake "the greatest adventure left to humankind," was voted down amid hoots of laughter from our fellow Americans in the U.S. Senate. Undaunted, scientists have now used private monies to start again. The Planetary Society, a 100,000-member organization founded by Carl Sagan, has sponsored a project called **BETA** (**B**illion-channel **ExtraT**errestrial **A**ssay). It has a receiver that can detect and search for signals at a quarter of a billion frequencies simultaneously using a 26-meter radio telescope belonging to Harvard University. The project has received widespread support from the scientific community as well as many notables such as the famous film director Steven Spielberg who has contributed substantial funding. The Planetary Society has also lent its support to another hunt for alien intelligence called **SERENDIP** (**S**earch for **E**xtraterrestrial **R**adio **E**missions from **N**earby **D**eveloped **I**ntelligent **P**opulations) at the University of California, Berkeley. And the doomed NASA effort has been revived, revised, and renamed ("**Project Phoenix**") by a group of scientists at the newly organized, privately funded **SETI** (**S**earch for **ExtraT**errestrial **I**ntelligence) **Institute** in California (Figure 6.9).

Are we alone in the Cosmos? We won't know until we do the experiments that can tell us. Let us hope the SETI Institute Phoenicians and their colleagues will be able to marshal the necessary resources to allow this grand adventure to continue until we have the answer.

Further Reading

The Quest for a Cosmic Connection.

The classic reference book for this entire subject is *Intelligent Life in the Universe* by I. S. Shklovskii and C. Sagan (1969; San Francisco: Holden Day)—some three decades old, it is still worth reading. Don Goldsmith and I have written a more modern text called *The Search for Life in the Universe* (2nd edition, 1992; Reading, MA; Addison Wesley). A series of articles on "Life in the Universe" by C. Sagan, S. J. Gould, M. Minsky, S. Weisberg, and others appeared in *Scientific American* **271**(4), October, 1994. An excellent collection of research papers covering many aspects of this subject appears in *Astronomical and Biochemical Origins and the*

FIGURE 6.9 Astronomer Jill Tarter (after whom the heroine of Carl Sagan's popular film *Contact* is modeled) at the console of the 140-foot-diameter radio telescope at Green Bank, West Virginia. In her hunt for signals of intelligent alien life, Dr. Tarter is peforming one of the experiments that has finally brought the subject of extraterrestrial civilizations into the domain of science. (Courtesy of Seth Shostak and the SETI Institute.)

Search for Life in the Universe, C. B. Cosmovici, S. Bowyer and D. Wertheimer, editors (1997; Bologna: Editrice Compositori).

The Myth of UFOs

The best serious discussion of this contentious topic is *UFOs Explained* by Philip J. Klass (1974; New York: Random House)—Klass, an editor for *Aviation Week and Space Technology Magazine* presents a refreshing, illuminating discussion of a large number of UFO reports of all types. In contrast, *The UFO Experience, A Scientific Inquiry* by J. Allen Hynek (1972; Chicago: Henry Regnery) is not what its subtitle suggests, despite the fact that its author was a professor of astrophysics at Northwestern University. What Hynek does accomplish is a classification of UFO reports, giving us the phrase "close encounters of the third kind," the title of a popular movie. *Flying Saucers, A Modern Myth of Things Seen in the Skies* by C. G. Jung (1978; Princeton, NJ: Bollingen Series) goes into the mythological aspect of UFOs from the perspective of the famous psychoanalyst.

The Myth of Ancient Astronauts

Chariots of the Gods? (1980) by Erich van Däniken is available from the nonfiction division of Berkeley Books, NY. The best antidote for this book is your own common sense, augmented by references available in any public library.

The Myth of Malevolent Martians

David Morrison and I have given a description of modern views of Mars in our book *The Planetary System* (2nd edition, 1995, Reading, MA: Addison Wesley). A large number of first-rate technical papers on various aspects of this enchanting planet are collected in the monumental volume *Mars*, edited by H. Kieffer, B. Jakosky, C. Snyder, and M. Matthews (1992; Tucson, AZ: University of Arizona). The serious student should also consult *The Surface of Mars* by Michael Carr (1981; New Haven, CT: Yale University Press) and *Water on Mars* (1997) by the same author and publisher.

The Myth of the Garden of Eden

David Morrison and I present a comparison of the inner planets in our book cited above. I have developed a model for cometary delivery of volatiles with Akiva Bar-Nun in *Icarus* **116**: 215-226 (1995) and in the *Proceedings of a Conference on Volatiles in the Earth and Solar System* (1995; AIP Conference Proceedings No. 341, edited by K. Farley, pages 123-138). I also review this subject in a paper called "Mars: Was there an Ancient Eden?", pages 203-218 in *Astronomical and Biochemical Origins and the Search for Life in the Universe*, edited by C. Cosmovici, S. Bowyer, and D. Werthimer (1997; Bologna: Editrice Compositori). The analysis of possible relic biological activity in the famous rock from Mars is clearly described in Chapter 5 of this volume and by Donald Goldsmith in *The Hunt for Life on Mars* (1997; New York: Dutton).

The Number of Civilizations in the Galaxy

Frank Drake's wonderful equation has been discussed and evaluated by many authors in many places. Readers are encouraged to substitute their own values for the various terms and calculate "N." Both "Shlovskii and Sagan" and "Goldsmith and Owen" (cited above) provide some guidelines. *Worlds Unnumbered, The Search for Extrasolar Planets* by Donald Goldsmith (1997; Sausalito, CA: University Science Books) gives a lucid, entertaining account of the new planets discovered in recent years.

Beating Myths into Science

The discussion in the present chapter is based in part on "Goldsmith-Owen" (cited above). Information about Project Phoenix is available from Dr. Jill Tarter at the SETI Institute, 2035 Landings Drive, Mountain View, CA 94043. Another excellent resource is The Planetary Society, 65 N. Catalina Avenue, Pasadena, CA 91106, which can provide information about the SERENDIP and BETA Projects.

GLOSSARY

A

Abd-A (abdominal A) **gene.** A gene present in the *Antennapedia* complex of homeotic genes in the fruit fly, *Drosophila.*

Active Mass Media. Communications media such as books, newspapers, magazines, tabloids, and the Internet, with which the user must be actively involved to absorb information.

Additive component. In analyses of heritability, that portion of the genetic component passed to offspring.

AIDS. Abbreviation for the human disease, Acquired ImmunoDeficiency Syndrome.

Ammonite. Any of numerous flat spiral fossil shells of cephalopods of the zoological order Ammonoidea, abundant especially in the Mesozoic.

Antibody. A protein produced by the immune system that recognizes a specific foreign antigen (often a toxic chemical) and triggers the immune response which attempts to destroy the antigen.

Antiscience. Theories, assumptions, and methods that ignore and are contrary to scientific facts and reasoning.

Artificial selection. Selective evolutionary pressure imposed by humans, for example by breeding cows for increased milk production.

Asteroid. One of thousands of small planetlike bodies orbiting between Mars and Jupiter with diameters from a fraction of a km to about 1,000 km.

Autapomorphy. Unique derived characteristic; a trait present in just one member of a lineage or in only one lineage among many.

Authigenic. Pertaining to a sedimentary mineral formed in the place it occurs.

B

BETA. Abbreviation for the Billion-channel ExtraTerrestrial Assay, a systematic scientific search for extraterrestrial intelligence sponsored by The Planetary Society.

Biological compound. Chemical compound produced by a living system, composed commonly of carbon, hydrogen, oxygen, nitrogen, sulfur, and/or phosphorus.

Biostratigraphy. The use of fossils or suites of fossils to determine the temporal relations among rock strata.

Bipedal. Two-footed posture and locomotion, with reference especially to the human attribute of upright stance on the hindlimbs.

Broca's area. A cortical region of the human brain located on the side of the frontal lobe, just above the temporal lobe (directly beneath a finger placed at the temple), known to be associated with the motor control of speech because injury to it results in asphasias (language disfunction).

Bryozoan. Any of a zoological group (Bryozoa) of small aquatic invertebrate animals that usually form permanently attached branched or mosslike colonies.

C

Cabinet. A collection of specimens, especially of mineralogical, biological, or numismatic interest.

Cambrian Explosion. The name given to the sudden appearance of the major animal phyla in the fossil record during the Cambrian Period.

Cambrian Period. The oldest geologic period of the Phanerozoic Eon of Earth's history, extending from approximately 543 to approximately 495 million years ago.

Carbon-14 (^{14}C). A radioactive isotope of carbon produced in the upper atmosphere and present in living plants and animals that can be used in carbon-14 dating because it decays to nitrogen (^{14}N) and a beta particle with a half-life of about 5,730 years.

Carbonate. Any of various minerals containing the chemical group CO_3^-, such as calcite ($CaCO_3$) or dolomite ($CaMg[CO_3]_2$), or a rock consisting chiefly of such minerals, such as limetone or dolostone.

Chordate. Any of various animals such as tunicates and vertebrates belonging to the phylum Chordata characterized by the presence of a dorsal notochord at some stage of development and a dorsal hollow nerve chord.

Chronostratigraphy. Stratigraphy based on correlation of index horizons, such as layers of volcanic ash, that serve as time markers.

Closure temperature. The temperature in a cooling magma below which diffusion of isotopes into or out of minerals effectively ceases marking the time dated by isotopic methods of geochronology.

Cnidaria. The phylum of animals that includes groups such as jellyfish, sea anemones, and corals.

Coccolith. Any of diverse types of photosynthetic and generally spheroidal and carbonate test-enclosed marine plankton of the group Coccolithophoridae.

Coelacanth. Any of a zoological family of largely extinct lobe-finned fish such as the living genus *Latimeria*.

Condylarth. Any of the particular group of small extinct Tertiary mammals typified by Eocene *Hyopsodus*.

Conformable geologic strata. Pertaining to a series of rock units that make up a structurally contiguous, vertically unbroken sequence.

Contingency. Pertaining to an event that is dependent on or conditioned by something else such as a prior event.

Continuous trait. A biological characteristic that evolves by a change in size, such as limbs, crania, and teeth in the primate lineage.

Cooling temperature. The temperature at which magnetic minerals align parallel to the Earth's magnetic field as an igenous rock cools and crystallizes from magma.

Correlatable geologic strata. Rock units in one set or series that can be shown to have been deposited at the same time as members of another set or series.

Cranial capacity. The internal volume of a skull.

Cretaceous Period. The youngest of the three geologic periods of the Mesozoic Era of Earth's history, extending from approximately 145 to 65 million years ago.

Critical thinking. A way to rationally evaluate claims and situations, based the abilities to (1) ask relevant questions, (2) define problems, (3) examine evidence, (4) analyze assumptions and biases, (5) avoid emotional reasoning, (6) avoid oversimplification, (7) consider alternate interpretations, and (8) tolerate uncertainty.

Crustacean. Any of a large zoological class (Crustacea) of mostly aquatic arthropods that have a chitinous or chitinous and calcareous exoskeleton and a pair of often much modified appendages on each body segment, such as lobsters, shrimps, and crabs.

Crustal plate. Thick oceanic or continental rocky masses that according to the global plate tectonic model move slowly across the Earth's surface propelled by movement of underlying material of the planetary interior.

Cyanobacterium. Any of a diverse group (Cyanobacteria) of prokaryotic microorganisms capable of oxygen-producing photosynthesis; known also as blue-green algae.

D

Darwin (*d*). A unit of measure of the rate of evolution, one *d* being equal to the rate of evolutionary change in size by a factor of *e* (the base of natural logarithms, = 2.718281828...) per million years.

Darwinian struggle. The competition involved in natural selection.

Developmental gene. A gene involved in the development of an organism (that is, from egg to adult).

Devonian Period. A geological period of the Phanerozoic Eon of Earth's history, extending from approximately 410 to approximately 360 million years ago.

Diatom. Any of a taxonomic group of unicellular algae occurring in marine or fresh water, each having a cell wall made of two halves impregnated with silica.

Dike. A tabular body of igneous rock that has been injected while molten into a fissure.

Diluvialist. An adherent of the view that the Biblical story of Noah and the Flood is historically accurate.

Diploblast. Any of various "lower invertebrates," such as jellyfish, that are composed of two tissue layers (an ectoderm and an endoderm) and so lack the third layer (the mesoderm) present in vertebrates and "higher invertebrates."

DNA (deoxyribonucleic acid). The genetic information-containing molecule of cells, a double-stranded nucleic acid made up of nucleotides that contain a nitrogenous base (adenine, guanine, thymine, or cytosine), deoxyribose sugar, and a phosphate (PO_4) group.

Down's syndrome. "Mongolism," a human genetic disorder first documented in detail by the English physician J. L. H. Down in 1896.

Drake Equation. A formula ($N = R* \cdot f_p \cdot n^e \cdot f_1 \cdot f_i \cdot f_c \cdot L$) developed by astronomer Frank Drake to estimate "N," the number of advanced civilizations in our galaxy at a given time, in which "$R*$" is the rate of star formation in the galaxy; "f_p," the fraction of stars that have planets; "n^e," the number of Earthlike planets in such planetary systems; "f_1," the fraction of such planets on which life develops; "f_i," the fraction of life-bearing planets that produce intelligent life; "f_c," the fraction of intelligent life-bearing planets that produce a civilization capable of and interested in interstellar communication; and "L," the average lifetime of such civilizations, measured in years.

E

East African Rift. A prominent region of geologic rifting, the result of movement of two continental tectonic plates, in central East Africa.

Eccentricity. Deviation of an orbit from a circular path.

Ediacaran Fauna. A Vendian-age assemblage of soft-bodied multicellular animals, the oldest fauna known.

Entropy. A measure of the disorder within a closed thermodynamic system.

Eocene Epoch. A temporal subdivision (epoch) of the Tertiary Period of Earth's history, extending from approximately 58 to approximately 37 million years ago.

Ets2 **gene.** A gene in vertebrates (in humans on chromosome 21, in mice on chromosome 16) that governs

Ets2 gene. *(continued)*
important aspects of the formation of cartilage including skull precursor cells.

Eukaryote. Any of a taxonomic group (Eucarya) of organisms composed of one or more nucleus-containing cells that constitutes one of three main branches of the Tree of Life (the other two, the Archaea and Bacteria, being composed of prokaryotic microorganisms).

Eurypterid. Any of the aquatic, usually large, Paleozoic-age arthropods included in the extinct zoological family Eurypterida.

Evidential reasoning. A way to evaluate the veracity of a claim, based on (1) honest assessment and (2) replicability of supporting evidence and the claim's (3) falsifiability, (4) underlying logic, (5) consistency with available evidence, and (6) intellectual sufficiency.

F

Facies. The overall characteristics of sedimentary strata that reflect their environment of deposition

Fault. A planar fracture in geologic strata accompanied by a displacement on one side of the fracture with respect to the other and in a direction parallel to the fracture.

Feedback loop. A circulating system in which a part of a system's output is returned to its input.

Foraminifera. Any of a protozoan order (Foraminifera) of marine rhizopods usually having calcareous shells perforated with minute holes for protrusion of slender psuedopodia.

Fractal. Pertaining to self-similarity of physical organization at two or more levels or degrees of observation.

G

Gene. A segment of DNA carrying information for production of a protein or RNA (ribonucleic acid) molecule.

Genetics. The science that deals with heredity and variation.

Genus. In biologicaal classification, a major category ranking above the species and below the family.

Geochronology. The scientific study of the chronology of the past as indicated by geologic data.

Geologic period. A formal division of geologic time longer than an epoch and included in an era.

Gradualism. In both geology and evolution, the concept of slow, gradual (rather than rapid, catastrophic) change.

Great ape. Any of several types of large semierect primates, the chimpanzees (including the bonobos), gorillas, and orangutans.

Greenhouse effect. Warming of the Earth's surface and the lower layers of the atmosphere caused by interaction of solar radiation with atmospheric gases (principally carbon dioxide, methane, and water vapor) and its conversion to heat.

H

Half life. The time required for half of something to undergo a process, such as the time required for half of the atoms of an amount of a radioactive isotope to be transformed to daughter atoms.

Haltere. The balancing organ apparatus (highly modified wings) of advanced files such as *Drosophila*.

Heredity. Transmission of genetic factors that determine individual characteristics from one generation to the next.

Heritability. The ability of genetically determined factors to be transmitted from one generation to the next.

HIV. Abbreviation for Human Immunodeficiency Virus, the cause of AIDS (acquired immunodeficiency syndrome).

HOM/Hox gene. Any of a family of homologous homeotic genes in vertebrate animals and fruit flies.

Homeotic mutation. A mutation that causes malformation in developing or regenerating animal tissue in which a segment or region of the body is transformed into the likeness of some other normal segment or region.

Hominid. Any member of the Hominidae, the zoological family that includes humans

Hoxd-11 gene. A gene in vertebrate animals that instructs particular cells to occupy appropriate locations in a developing body.

Hydrologic cycle. The cyclic circulation of water from the surface of the land, to the soil and underlying rock, to streams, rivers, and the ocean, and then into the atmosphere and back to the surface of the land.

Hyperthermophile. Any of various prokaryotic microorganisms, such as diverse members of the Archaea, that survive and grow in exceptionally high temperature (>80° C) environments.

I

Ichthyosaur. Any of an extinct group (Ichthysauria) of marine fishlike or porpoiselike reptiles abundant in Mesozoic seas.

Iconography. Pictorial material relating to or illustrating a subject.

Igneous. Rock formed from solidification of molten magma.

Index fossil. Fossils that typically are easily recognizable, exist over a relatively short period of geologic time, and are widespread and so are useful to data rock strata.

Influenza. A highly contagious virus disease in humans and other animals marked by respiratory symptoms.

Invertebrate. Any of diverse animals that lack backbones; "lower metazoans."

Isochron. A parameter experimentally determined from comparison of the isotopic compositions of two or more components (usually minerals) that share a common age and used in isotopic dating of geologic materials.

Isotope. Any of two or more types of atoms of a chemical element that have nearly identical chemical behavior but differing atomic mass and physical properties. Some isotopes are stable whereas others decay radioactively.

J

Jurassic Period. The second oldest of the three geologic periods of the Mesozoic Era of Earth's history, extending from approximately 210 to approximately 145 million years ago.

L

Law of Superposition. The geologic rule that in a sequence of undeformed strata the oldest rock layer is at the bottom and each layer upward is progressively younger.

Lithostratigraphy. Correlation of geologic strata by use of the characteristics of specific rock types.

Lobe-finned fish. Any of a taxonomic group of fish that have both bone and muscle in their limbs, in contrast to having simple fins as in most bony (teleost) fish; represented by seven living species, the coelacanth and six species of lungfish.

Lungfish. Any of six living and numerous fossil species that together with the coelacanth are classified as lobe-finned fish.

M

Ma. Milliard anna, one million (1×10^6) years.

Magma. Molten rock such as volcanic lava.

Magnetite. A mineral, the iron oxide Fe_3O_4.

Magnetofossil. Term used by some scientists to refer to minute grains of magnetite which they regard as possibly biologic (bacterial) in origin.

Mass media. Widely available communications media, both those that require active involvement by the consumer (books, newspapers, magazines, tabloids, the Internet) and a more passive participation (such as television, radio, and movies).

Maxilliped. Any of the crustacean appendages that comprise the first pair or first three pairs situated next behind the maxillae, a specific set of mouthparts.

Meristic trait. A biological characteristic that evolves by a change in number, in animals such as body segments, toes, and neck vertebrae.

Mesozoic Era. The second oldest of three geologic eras that make up the Phanerozoic Eon of Earth's history, extending from approximately 251 to 65 million years ago.

Metamorphism. A change in the constitution of a rock produced by pressure and heat.

Milankovitch cycle. Any of the series of cyclic changes in climate proposed by astronomer Milutin Milankovitch to relate to orbital dynamics including the combined effects of periodic variations in the tilt of the Earth's axis, the wobble (precession) of the axis, and the ellipticity of the orbit.

Missing link. An absent member needed to complete a series, in paleoanthropology a hypothetical intermediate between humans and simian progenitors.

Morphology. The form and structure of an organism or any of its parts.

Mycoplasma. Any of a genus (*Mycoplasma*) of extremely minute parasitic microorganisms that lack cell walls; known also as PPLO, *Pleuropneumonia*-like organisms.

N

Natural logarithm. A system in mathematics for which *e* (2.718281828...)

is the base; that is, the natural logarithm of *e* is unity.

Natural selection. Preferential survival of individuals having advantageous variations relative to other members of their population or species. For natural selection to operate, competition for resources (a struggle for survival) and suitable variation among individuals must exist.

Neomorphic trait. A biological characteristic that significantly differs in form or structure from its evolutionary precursor, such as bird feathers derived from the epidermal scales of repitilian ancestors.

Neurophysiology. The physiology of the nervous system.

Noble gas. Any of a group of rare gases that includes helium, neon, argon, krypton, and xenon and exhibit great stability and extremely low reaction rates; known also as inert gas.

Nonadditive component. In analyses of heritability, that portion of the genetic component not passed to offspring.

Notochord. A characteristic of chordate animals: A stiff phosphatic rod that functions to preserve body shape during locomotion in some species.

Nymph. Any of various immature insects, especially a larva of a dragonfly or mayfly that is incompletely developed.

O

Order, taxonomic. See *Taxonomic order.*

Ordovician Period. A geological period of the Phanerozoic Eon of Earth's history, extending from approximately 495 to approximately 440 million years ago.

Organelle. A specialized membrane-enclosed structrure in a cell, such as a mitochondrion or a plastid, that performs a specific function.

Organic compound. A chemical compound of the type typical of, but not restricted to, living systems, composed commonly of carbon, hydrogen, oxygen, nitrogen, sulfur, and/or phosphorus.

Oxide. A chemical compound of oxygen and another element as in the minerals hematite, Fe_2O_3, and magnetite, Fe_3O_4.

P

PAH. See *Polycylic aromatic hydrocarbon.*

Passive Mass Media. Communications media such as television, radio, and movies, which require minimal active participation by the user.

Peer review. A scrutiny of reported findings for accuracy and completeness by fellow scientists.

Permian Period. A geological period of the Phanerozoic Eon of Earth's history, extending from approximately 290 to approximately 251 million years ago.

Phanerozoic Eon. The younger of two principal divisions (eons) of Earth's history, extending from the beginning of the Cambrian Period, approximately 543 million years ago, to the present; the Phanerozoic and the older Precambrian Eon include all geologic time.

Phyletic gradualism. A model of evolutionary mode characterized by slow more or less steady modification of biologcal structures over long periods of geologic time.

Phylogeny. The evolutionary relations among a group of organisms.

Plate, geological. See *Crustal plate.*

Plate tectonics. Global tectonics based on an Earth model characterized by many thick oceanic or continental plates that move slowly across the global surface propelled by movement of underlying material of the planetary interior.

Plesiosaur. Any of a zoological group (Plesiosauria) of extinct marine Mesozoic reptiles.

Pluton. A mass of igneous rock solidified at depth.

Polycyclic aromatic hydrocarbon (PAH). Any of various organic compounds composed of a few to many six-membered rings of carbon atoms linked by an alternating series of single and double bonds and to which are attached hydrogen atoms.

Population, species. A group of organisms of the same species that occupy a defineable geographic region and have reproductive continuity from generation to generation.

Precambrian Eon. The older of two principal divisions (eons) of Earth's history, extending from the formation of the planet, approximately 4,500 million years ago, to the beginning of the Cambrian Period, approximately 543 million years ago; the Precambrian and the younger Phanerozoic Eon include all geologic time.

Precession. The wobble of the axis of a rotating planet.

Principle of Mediocrity. A term coined by I. S. Shklovskii and C. Sagan to express the assumption that human society represents an average example of an advanced civilization in our galaxy.

Principle of Uniformitarianism. The concept that processes and natural laws in operation today have not changed over Earth's history; sometimes phrased as "the present is the key to the past."

Project Phoenix. A systematic scientific search for signals from extraterrestrial civilizations conducted by the SETI Institute.

Prokaryote. Any of diverse types of non-nucleated microorganisms of the Archaea and Bacteria.

Protein. A polymeric organic compound composed of a string of amino acid monomers.

Pseudoscience. Theories, assumptions, and methods erroneously regarded as scientific.

Pterosaur. Any of a zoological group (Pterosauria) of extinct Mesozoic flying reptiles.

Punctuated equilibrium. A model of evolutionary mode characterized by relatively rapid modification of biological structures over short intervals of geologic time, associated especially with speciation, interspersed within long periods of relatively little evolutionary change (stasis).

R

Radioactivity. The property possessed by some elements (such as uranium) of spontaneously emitting alpha or beta rays by the disintegration of the nuclei of atoms.

Radiolarian. Any of a large protozoan order (Radiolaria) of marine plankton having a siliceous shell and radiating threadlike pseudopodia.

Ray-finned fish. See *Teleost*.

Relativistic. Moving at such high velocity that there is a significant change in properties (such as mass) in accordance with the theory of relativity.

Response to selection. The outcome of natural or artificial selection of one or more biologic traits in a population.

Rock cycle. The cyclic circulation of rock materials from a sedimentary deposit, to a metamorphic rock, an igneous rock, and back again to become constituents of sediments.

S

Sagittal crest. A bony protuberance situated above the suture between the parietal bones of a skull.

Science illiteracy. Insufficient understanding and knowledge of the facts and methods of science.

Scientific method. A formalized statement of the principles and procedures for the systematic pursuit of knowledge involving (1) recognition and formulation of a problem, (2) the collection of data through observation and experiment, and (3) the formula-

tion and testing of hypotheses.

Scientifically literate. The property of having a working knowledge of scientific facts and sufficient understanding of the methods of science such as the abilities to think critically, use evidential reasoning to draw conclusions, and evaluate scientific authority.

Second Law of Thermodynamics. A physical law stating that systems closed to the exchange of matter and energy with their surroundings inevitably progress toward equilibrium with an increase in disorder.

Sedimentary. Particulate commonly granular mineralic material deposited by water, wind, or glaciers that on compression can be lithified to a sedimentary rock.

Selection differential. The strength, degree of influence, of selection for or against a particular biological trait.

SERENDIP. Abbreviation of the Search for Extraterrestrial Radio Emissions from Nearby Developed Intelligent Populations, a systematic scientific hunt for extraterrestrial intelligence conducted at the University of California, Berkeley.

SETI Institute. The Search for ExtraTerrestrial Intelligence Institute situated in northern California.

Speciation event. The process of origination of a biological species.

Species. The fundamental category of biological classification, ranking below the genus and in some species composed of subspecies or varieties. Among various definitions, the most common is the *Biological Species Concept*: "Species are actually or potentially interbreeding natural populations which are reproductively isolated from other such groups."

Spinal cord. The longitudinal chord of nervous tissue extending from the brain along the back in the spinal canal.

Stabilizing selection. Natural selection against deviations from the average of a population (and, so, against evolutionary change), such as the strong selective pressure against both unusually low and high birth weights in humans.

Stasis. In the punctuated equilibrium model of evolution, the relatively long periods characterized by little evolutionary change.

Sternum. The breastbone in vertebrates.

Stratigraphy. The study of sedimentary beds (strata) laterally and vertically, including all aspects of the sedimentary materials—their mineralogy, physcial characteristics, origin, depositional environment, fossil components, and so forth.

Stromatolite. Accretionary organosedimentary structure, commonly finely layered, megascopic, and calcareous, produced by the activities of mat-building microorganisms, principally filamentous photosynthetic prokaryotes such as various types of cyanobacteria.

Sulfide. A chemical compound of sulfur and another element as in the mineral pyrite, FeS_2.

Synapomorphy. A shared derived characteristic used to identify lines of descent in the Tree of Life; an evolutionary innovation shared by members of a lineage but not present in members of any other lineage.

T

Taxonomic order. In biological classification, a major category ranking above the family and below the class.

Taxonomy. The theory and practice of classifying organisms.

Tectonic. Pertaining to a crustdeforming process or event.

Teleost. Any of a group (Teleostei) of fishes characterized by a bony skeleton (rather than a cartilaginous one as in sharks and rays); known also as ray-finned fish.

Tertiary Period. The older of two geologic periods of the Cenozoic Era of Earth's history, extending from 65 to 1.6 million years ago.

Tetrapod. A vertebrate animal (such as a frog, bird, or cat) having two pairs of limbs.

Thermodynamics. A statistical branch of physics concerned with the conversion of one form of energy into another.

Theropod. Any of a zoological subgroup (Theropoda) of obligately bipedal saurischian reptiles that includes all carnivorous dinosaurs such as *Deinonychus*, *Velociraptor*, and *Tyrannosaurus*.

Thoracic. Pertaining to the throax, the upper back region.

Tree of Life. A branching, treelike representation showing the relatedness of all living organisms, commonly based on comparison of rRNAs, the ribonucleic acids of their protein-manufacturing ribosomes.

Triassic Period. The oldest of three geologic periods of the Mesozoic Era of Earth's history, extending from approximately 251 to approximately 210 million years ago.

Trilobite. Any of diverse types of extinct arthropod animals of the Paleozoic Era (543 to 245 million years ago) characterized by a three-lobed body organization.

Tuff. A rock composed of fine volcanic detritus, usually fused together by heat.

U

***Ultrabithorax* (*Ubx*) gene.** A gene in the homeotic *Bithorax* complex of the fruit fly, *Drosophila*.

Unconformity, geological. A surface separating two series of rocks that represents a period when deposition did not occur; a gap in the rock record.

V

Varve. A pair of layers of alternately finer and coarser silt or clay that comprise an annual cycle of deposition in a body of quiet water.

Vendian Period. The youngest geological period of the Precambrian Eon of Earth's history, extending from approximately 620 to approximately 543 million years ago.

Vertebra. One of the bony segments composing the spinal column in vertebrate animals.

Vertebral foramen. The central hole of a vertebra that encloses the spinal chord.

Vertebrate. Any member of the Vertebrata (a subphylum of the Chordata) that consists of all animals that possess a bony or cartilaginous skeleton and a well-developed brain such as fishes, amphibians, reptiles, birds, and mammals.

Virus. An ultramicroscopic, intracellular complex of nucleic acid (DNA or RNA) and protein incapable of autonomous replication or metabolism.

Z

Zircon. A mineral, $ZrSiO_4$.

INDEX